e-Health Security Management

SCIENCES

Networks and Communications,
Field Director – Guy Pujolle

Network Management and Control,
Subject Head – Francine Krief

e-Health Security Management

Communications Security, Data Processing Security and Patient Privacy

Coordinated by
Omessaad Hamdi

iSTE

WILEY

First published 2024 in Great Britain and the United States by ISTE Ltd and John Wiley & Sons, Inc.

ISTE Ltd
27-37 St George's Road
London SW19 4EU
UK

www.iste.co.uk

John Wiley & Sons, Inc.
111 River Street
Hoboken, NJ 07030
USA

www.wiley.com

Library of Congress Control Number: 2024942248

British Library Cataloguing-in-Publication Data
A CIP record for this book is available from the British Library
ISBN 978-1-78945-179-5

ERC code:
PE6 Computer Science and Informatics
PE6_5 Cryptology, security, privacy, quantum cryptography
PE6_7 Artificial intelligence, intelligent systems, multi agent systems
PE7 Systems and Communication Engineering
PE7_8 Networks (communication networks, sensor networks, networks of robots, etc.)
PE7_11 Components and systems for applications (in e.g. medicine, biology, environment)

Contents

Chapter 1. Overview of e-Health Architectures 1
Omessaad HAMDI

1.1. Introduction . . . 1
1.2. Definitions . . . 2
1.2.1. e-Health . . . 2
1.2.2. Telehealth . . . 2
1.2.3. m-Health . . . 2
1.2.4. Telemedicine . . . 2
1.3. e-Health services . . . 3
1.4. Requirements for e-health systems . . . 4
1.5. e-Health system architecture . . . 5
1.5.1. Components of an e-health architecture . . . 6
1.5.2. Features of e-health systems . . . 6
1.6. e-Health system technologies . . . 8
1.6.1. Devices . . . 8
1.6.2. Connecting technologies . . . 9
1.6.3. Other technologies . . . 10
1.7. Security in e-health systems . . . 12
1.7.1. Security services . . . 12
1.7.2. Legal environment for e-health systems . . . 13
1.8. Medical data security . . . 14
1.8.1. Cryptography . . . 14

1.8.2. Biometrics 16
1.8.3. Blockchain 17
1.9. Perspectives 19
1.10. Conclusion 20
1.11. References 21

Chapter 2. Vulnerabilities in e-Health and Countermeasures 27
Aida BEN CHEHIDA DOUSS and Ryma ABASSI

2.1. Introduction 27
2.2. The importance of digitization in healthcare systems 28
2.3. The challenges of digitization in e-health systems 30
2.4. Cyber-attacks in the healthcare sector 31
2.4.1. Profiles of cybercriminals 32
2.4.2. Motivations of cybercriminals 33
2.4.3. Risks and repercussions 35
2.4.4. Types of attacks 36
2.5. Security incidents in the healthcare sector 39
2.5.1. Example of a phishing attack 40
2.5.2. Examples of ransomware attacks 40
2.5.3. Examples of data theft attacks 41
2.5.4. Examples of DDoS attacks 42
2.5.5. Example of an internal attack 42
2.6. Existing security measures for e-health systems 42
2.7. Recommendations for protecting e-health systems 45
2.7.1. Risk management methods 45
2.7.2. Technical and organizational recommendations 46
2.7.3. Raising awareness and training 47
2.8. Conclusion 48
2.9. References 49

Chapter 3. Security Policies for e-Health Systems 53
Ryma ABASSI

3.1. Introduction 53
3.2. The concept of the security policy 54
3.2.1. Definition 54

3.2.2. Modeling a security policy 57
3.3. Environment for specifying, validating and testing security policies 61
3.3.1. Specifying a security policy 61
3.3.2. The concept of executable security policy 63
3.3.3. Testing a security policy 64
3.4. Security services for e-health systems 66
3.4.1. The e-health concept 66
3.4.2. Comparison of national digital health infrastructure security policies 67
3.5. Security requirements for e-health platforms 69
3.5.1. Essential security functions. 69
3.5.2. Security models 70
3.6. Future security challenges for e-health 73
3.7. Conclusion 74
3.8. References. 74

Chapter 4. Adaptive, Dynamic, Decentralized Authorizations for e-Health 77
Tidiane SYLLA, Mohamed AYMEN CHALOUF, Léo MENDIBOURE and Francine KRIEF

4.1. Introduction. 77
4.2. Fundamental principles 79
4.2.1. Concept of e-health. 79
4.2.2. Context-aware computing and security in the IoT 81
4.2.3. Authentication and Authorization for Constrained Environments (ACE-OAuth) 86
4.2.4. Blockchain 89
4.3. Proposal for dynamic, decentralized adaptation of e-health authorizations 91
4.3.1. Threat model for the environment under consideration . . 91
4.3.2. Proposed architecture for dynamic, decentralized authorization management 92
4.4. Conclusion 100
4.5. References. 101

Chapter 5. Applying Blockchain to e-Health 107
Cyrine LAHSINI, Faiza HAMDI and Omessaad HAMDI

5.1. Introduction 107
5.2. Blockchain technology 108
5.2.1. Blockchain fundamentals 108
5.2.2. Blockchain categories 110
5.2.3. Characteristics of the blockchain 112
5.3. Health sector 113
5.3.1. Patients 113
5.3.2. Doctors 114
5.3.3. Insurance sector 114
5.3.4. Pharmaceutical industry 115
5.3.5. Government 115
5.4. Issues and challenges for the healthcare sector 115
5.4.1. Quality 116
5.4.2. Coordination 117
5.4.3. Integrity 117
5.4.4. Transparency 118
5.4.5. Traceability 118
5.4.6. Interoperability 119
5.4.7. Data sharing 120
5.4.8. Costs 120
5.4.9. Data volume 121
5.5. Application of blockchain technology in e-health systems . . . 122
5.5.1. Electronic health records 122
5.5.2. Pharmaceutical supply chain 123
5.5.3. Patient follow-up 124
5.5.4. Scientific research in the health sector 125
5.5.5. Analyzing medical data 126
5.6. Implementing blockchain technology in healthcare 127
5.6.1. MedRec 127
5.6.2. MedCredits 128
5.6.3. MIStore 128
5.6.4. Robomed 129
5.6.5. HealthChain 129
5.6.6. Medicalchain 129

5.7. Contribution of the blockchain solution 130
5.8. Conclusion 133
5.9. References. 134

Chapter 6. Using Biometrics to Secure Intra-BAN Communications. 137
Omessaad HAMDI, Mohamed AYMEN CHALOUF and Amal SAMMOUD

6.1. Introduction. 137
6.2. Security for WBAN. 138
6.2.1. Architecture of an e-health system 138
6.2.2. Security requirements for WBANs 139
6.2.3. WBAN attacks. 140
6.3. Security solutions for intra-WBAN communications 140
6.3.1. TinySec 141
6.3.2. Biometric methods 141
6.3.3. ZigBee security 141
6.3.4. Bluetooth security. 141
6.3.5. Elliptical curve cryptography 142
6.4. Biometric data-based security solutions for WBANs 143
6.4.1. Biometrics 143
6.4.2. Examples of security approaches for intra-WBAN communications using biometrics. 145
6.4.3. The approach of Sammoud et al. 147
6.5. Discussion. 154
6.6. Conclusion 155
6.7. References. 158

Chapter 7. Using Biometrics for Authentication in e-Health Systems 161
Omessaad HAMDI, Mohamed AYMEN CHALOUF and Amal SAMMOUD

7.1. Introduction. 161
7.2. e-Health systems. 162
7.2.1. Architecture 162
7.2.2. Security services 163

7.3. Authentication techniques 163
7.3.1. Authentication factors 164
7.3.2. Types of authentications 164
7.4. Biometric authentication 166
7.4.1. Biometric features 166
7.4.2. Biometric system effectiveness 167
7.4.3. Performance measures for biometric systems 168
7.5. Multimodal authentication 168
7.6. Multi-factor authentication approaches for e-health system security 169
7.6.1. Sammoud et al.'s approach 173
7.7. Conclusion 178
7.8. References 179

Chapter 8. Security of Medical Data Processing 183
Manel ABDELHEDI and Omessaad HAMDI

8.1. Introduction 183
8.2. Homomorphic encryption 185
8.2.1. Definition 185
8.2.2. Terminology 186
8.2.3. Partially homomorphic encryption 187
8.2.4. Somewhat homomorphic encryption 190
8.2.5. Fully homomorphic encryption 191
8.2.6. Comparative study 193
8.2.7. Application of HE to secure e-health solutions 198
8.3. Attribute-based encryption 200
8.3.1. Key-policy attribute-based encryption 201
8.3.2. Ciphertext-policy attribute-based encryption 202
8.3.3. Comparative study 203
8.3.4. Application of ABE to secure e-health solutions 204
8.4. Conclusion 206
8.5. References 207

Chapter 9. Artificial Intelligence for Security of e-Health Systems 213
Mohamed Aymen CHALOUF, Hana MEJRI and Omessaad HAMDI

9.1. Introduction 213
9.2. e-Health systems 214
9.3. e-Health system security 215
9.3.1. Potential attacks 216
9.3.2. Security services 216
9.3.3. Security solutions 218
9.4. Artificial intelligence techniques 220
9.4.1. Machine learning 221
9.4.2. Deep learning 222
9.5. Intrusion detection based on artificial intelligence 223
9.5.1. IDS based on supervised learning 224
9.5.2. IDS based on unsupervised learning 225
9.5.3. IDS based on deep learning 226
9.6. AI-based IDS in WBANs 226
9.6.1. Tested learning techniques 227
9.6.2. Implementation and results 227
9.7. Conclusion 232
9.8. References 233

List of Authors 237

Index 239

1

Overview of e-Health Architectures

Omessaad HAMDI
IEEE, Rennes, France

1.1. Introduction

Digitization occupies a central place in all our daily activities, and the healthcare field is particularly affected by this digital evolution, which has considerably improved patient care (Hermes et al. 2020; Gupta et al. 2021). This improvement is based on two key factors: the increased involvement of patients in the management of their health, and easy access for healthcare professionals to digital tools and services.

Digitization is also improving people's quality of life, in terms of well-being and autonomy, and is helping respond to the growing number of elderly people worldwide. The phenomenon of aging is becoming a growing concern. To enable this population to age in a secure environment with a good quality of life, while reducing costs, several approaches have been developed.

In this chapter, we focus on e-health architectures. We begin by introducing the terms used in e-health. Next, we present the services offered by e-health systems and their requirements. The final sections will focus on security and the techniques used to guarantee the required security services. Finally, we look ahead to the future of e-health.

e-Health Security Management,
coordinated by Omessaad HAMDI.

1.2. Definitions

1.2.1. *e-Health*

The term *e-health* refers to information and communication technologies (ICT) combined with the Internet in the service of health.

1.2.2. *Telehealth*

Telehealth is part of e-health. It refers to the use of tools for producing, transmitting, and managing digitized medical information. Telehealth encompasses telemedicine and mobile health (m-health).

1.2.3. *m-Health*

m-Health is part of telehealth. It refers to healthcare practices supported by mobile devices, such as cell phones, patient monitoring systems and other wireless devices. The term includes, among others, applications such as wellness apps. Bashshur et al. (2011) point out that m-health is the only ICT-based healthcare field that can be justified solely based on mobility.

1.2.4. *Telemedicine*

Telemedicine is part of telehealth. It refers to the digital transmission of medical information (images, recordings, etc.) for remote diagnosis, specialist advice and continuous monitoring of a patient.

There are four forms of telemedicine (2010 decree):

– Remote consultation is between a healthcare professional and a patient: It refers to the use of communication technologies to provide health consultations to patients in geographically different locations.

– Remote education is between healthcare professionals, in the absence of the patient: It consists of a remote request for advice from colleagues based on information provided by the patient.

– Remote monitoring involves remote monitoring of a patient's health parameters, providing assessments of the patient's state of health.

– Remote assistance occurs when a doctor remotely guides a medical act. This can take place between two healthcare professionals or between a healthcare professional and a third-party present with the patient, for example, in an emergency.

Figure 1.1 summarizes the components of telehealth.

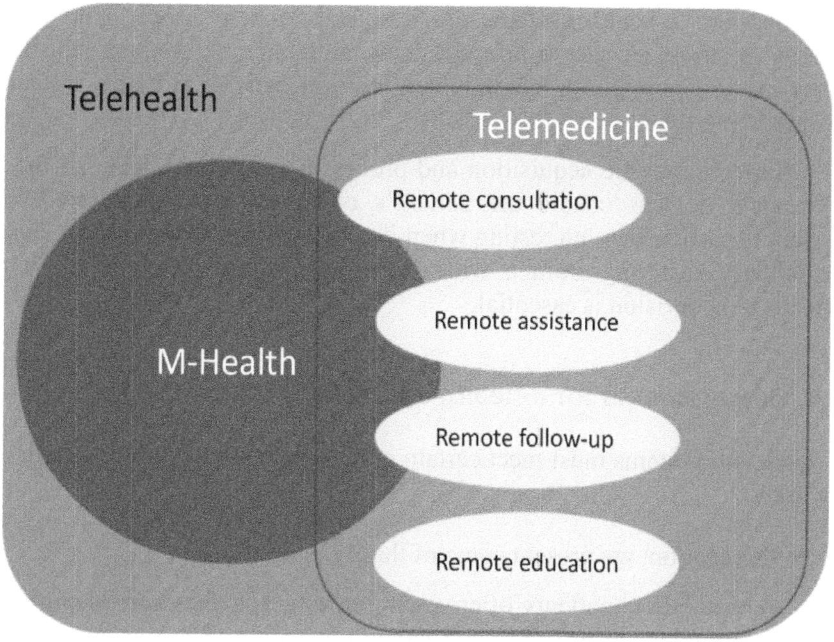

Figure 1.1. *Components of e-health*

1.3. e-Health services

e-Health offers a wide range of services designed to improve the quality of care and accessibility to medical services thanks to digital technologies:

– Cost reduction: e-health considerably reduces hospitalization and the need to keep elderly people in nursing homes. It also enables early detection of illness. Both services can significantly reduce healthcare costs (Atienza et al. 2007; Kostkova 2015).

– Social inclusion: the use of e-health technologies enables patients to remain active and independent as long as possible, enabling them to overcome their illness and/or disability without being excluded from society.

– Prevention: body and environmental data collected from sensors can be interpreted. By effectively managing these data, doctors can uncover facts and detect illness at an early stage.

– Support: e-health systems are designed to help people who are ill, elderly or disabled, and to promote their autonomy, safety and well-being. They make it possible to maintain and monitor patients at home, instead of hospitalizing them.

– Supervision: the acquisition and processing of patient data and the use of several devices enable the patient's condition to be monitored. This system is particularly interesting when it comes to high-risk patients, such as the elderly suffering from a wide range of chronic illnesses, for whom effective supervision is essential.

1.4. Requirements for e-health systems

e-Health systems must meet certain requirements if they are to be adapted by users.

In this section, we present some of these requirements:

– Acceptability: patients often wear sensors, and these are deployed in their environment to provide continuous monitoring. The sensors deployed must meet conditions of comfort and acceptability.

– Reliability: an e-health system must generate a very low false alarm rate.

– Energy autonomy: the energy autonomy of sensors plays an important role. Replacing sensor batteries is often complicated and/or costly.

– Ergonomics: it is essential that the devices and applications used are ergonomic and user-friendly to guarantee ease of use.

– Safety: devices and applications must comply with standards and regulatory requirements.

– Privacy protection: this is of paramount importance when dealing with media information, as this is sensitive data. To guarantee this protection, appropriate mechanisms must be put in place, especially in an environment where several users are involved.

Figure 1.2 summarizes the services and requirements of e-health systems.

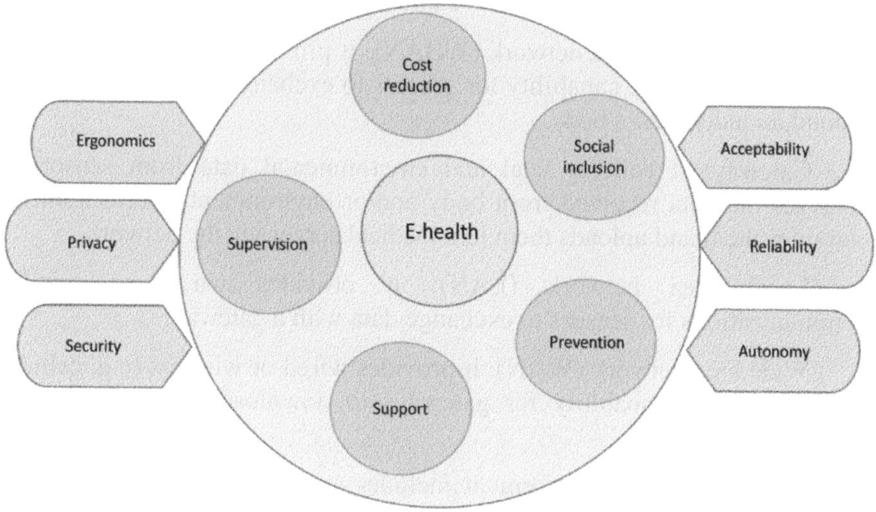

Figure 1.2. *Services and requirements for e-health systems*

1.5. e-Health system architecture

Different e-health system architectures have been developed to meet the specific needs of each project.

An architecture that summarizes most of the architectures proposed in the literature is shown in Figure 1.3.

In all e-health architectures, information flows from the patient to a medical server. Data are transferred from the sensors to a gateway that manages the sensors. Data transfer in the network can be continuous or ad hoc. Collected data are stored in a gateway, and then uploaded to a medical server.

1.5.1. *Components of an e-health architecture*

The main components of an e-health system are as follows (Hamdi et al. 2014):

– Sensors: these are devices that capture, store, process and transmit data.

– Wireless body area network (WBAN): it provides short-range wired or radio communication capability for sensors to exchange data with a gateway around an individual's body.

– Gateway: it collects vital and environmental data from sensors. It analyzes the data received from body and/or environmental measurements, compiles them and uploads them to a medical server via the network.

– Local area network (LAN): it provides wired or wireless communications for sensors to exchange data with a gateway.

– Wide area network (WAN): it provides wired or wireless (e.g. cellular) communications capability for gateways to download data to a medical server.

– e-Health systems platform: it includes servers for storing, processing and securing medical data.

Figure 1.3 gives an overview of the main components of e-health systems.

1.5.2. *Features of e-health systems*

– Data capture: this layer refers to the collection of patient data from vital signs and/or environmental sensors.

– Computation: this layer includes data analysis, management and personalization of care.

– Communication and storage: this layer covers vital signs communication, calculation and storage modules.

– Access: this layer refers to the way in which data are accessed. It often takes the form of a web portal or mobile application connected to a secure system hosted in the cloud, enabling continuous monitoring of patients' health status.

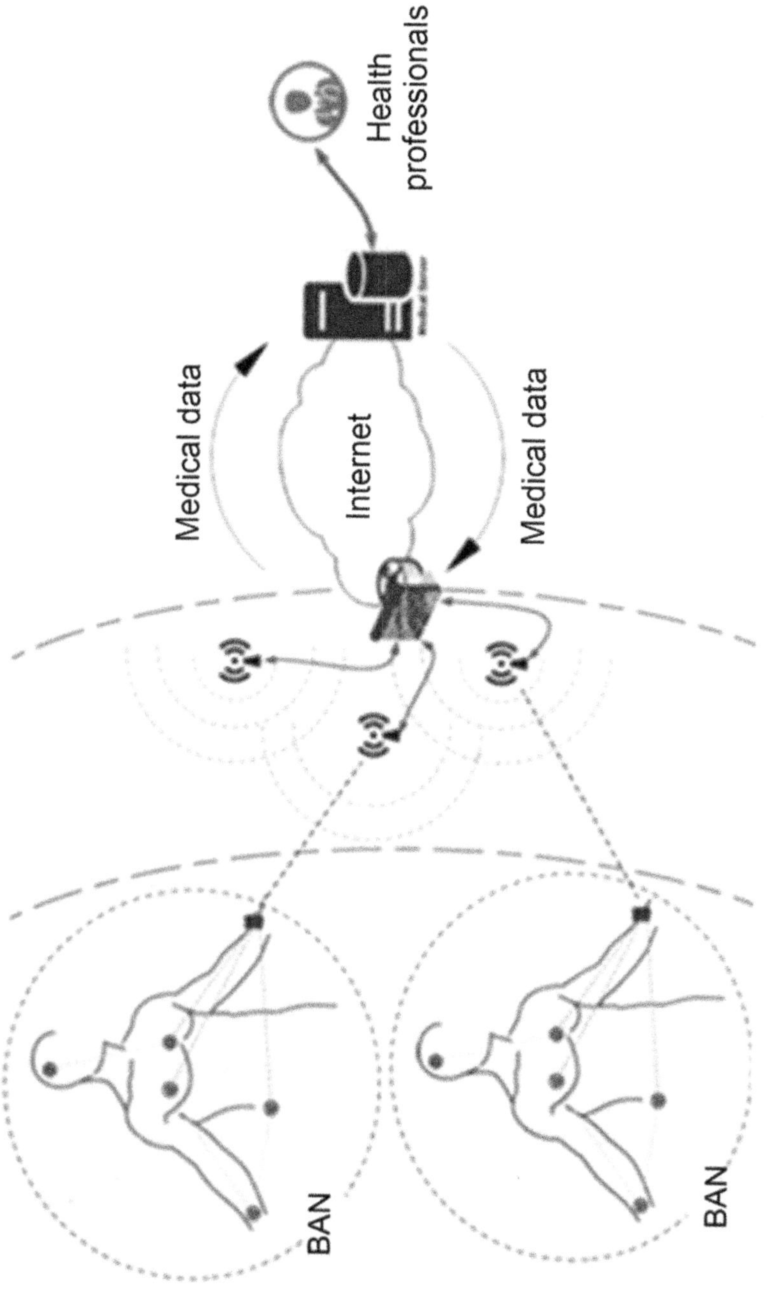

Figure 1.3. *Architecture of e-health systems (Hajar et al. 2021)*

1.6. e-Health system technologies

Connection technologies such as Bluetooth, WiFi, Internet and ZigBee play a key role in the growth of e-health applications and systems. When used in conjunction with other technologies, such as the Internet of Things (IoT), robotics, artificial intelligence (AI), cloud and Big Data, high-performance e-health systems can be created (Devedžić et al. 2021).

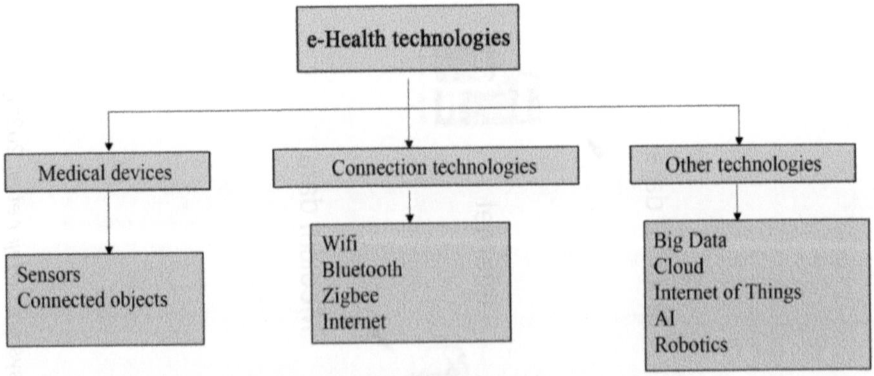

Figure 1.4. *e-Health technologies*

Figure 1.4 illustrates the main technologies used in e-health systems. These are grouped into medical devices, connection technologies and other technologies, and are detailed below.

1.6.1. *Devices*

Devices are mainly made up of sensors and connected objects, which play a key role in monitoring and ensuring the well-being of individuals, offering medical, safety and wellness services (Javaid et al. 2022).

– Sensors are devices that detect and measure specific information, such as body temperature, heart rate, blood pressure, physical activity, sleep quality and so on. These sensors collect valuable data on people's health and well-being.

– Connected objects, also known as IoT devices, are devices that can connect to the Internet and exchange data. These can include smartwatches, connected bracelets, connected scales, blood pressure monitors, thermometers and many more. These connected objects work in tandem with sensors to collect, transmit and analyze data relating to users' health and well-being (Fagroud et al. 2019; Balakrishnan et al. 2021).

Using these sensors and connected objects, medical services can provide precise monitoring of an individual's health status, detecting signs of potential health problems and enabling early intervention. Security services can use these devices to ensure the safety of the elderly or people at risk by detecting falls or monitoring unusual movements.

Several approaches have focused on the uses of sensors or connected objects in the medical field. Table 1.1 illustrates a few examples.

References	Sensor	Proposition
(Rabbani et al. 2021)	Implant	A real-time immune response monitoring system used in cancer therapy to track disease progression and provide personalized care.
(Gourob et al. 2021)	Artificial hand	Human–robot interactions to control a patient's hand gesture recognition system.
(Basaklar et al. 2021)	Smart clothing	A portable, low-energy device for personalized care without manual intervention.
(Gupta et al. 2021)	WBAN	A system for monitoring psychological parameters such as temperature and heart rate to provide real-time diagnosis.
(Hodgkiss and Djahel 2022)	WBAN	The use of biometric data to ensure strong authentication as part of an e-health system.
(Behera 2022)	Patches	Using chip-less RFID sensors to measure data and monitor vital signs in real time.

Table 1.1. *Application of sensors and connected objects in e-health systems*

1.6.2. *Connecting technologies*

In this section, we present the connection technologies used in the various components of an e-health architecture.

1.6.2.1. *ZigBee*

ZigBee is a wireless technology offering long battery life, low data rate and a secure network (Chung et al. 2013; Minakshi 2016). In addition,

ZigBee is an easy network to install and configure, supports various network topologies and allows for a large number of nodes to be connected. ZigBee meets the specific requirements of WBANs.

1.6.2.2. *Bluetooth*

Bluetooth was designed for short-range wireless communications, where several Bluetooth devices form a short-range network (Negra et al. 2016). Bluetooth is widely used in WBANs.

1.6.2.3. *LPWAN*

LPWAN technology, proprietary to the LoRa (Long Range) Alliance, consists of two main elements, LoRa and the LoRaWAN protocol. This technology has been the focus of much research into e-health systems, due to its low cost, long coverage area and long sensor lifetime (Sundaram et al. 2019).

1.6.3. *Other technologies*

In recent years, e-health has become more efficient and smarter thanks to cloud technologies, Big Data, AI and robotics.

1.6.3.1. *Big Data*

The application of Big Data in the e-health sector has enabled the better exploitation of data to diagnose disease and improve quality of care.

Online e-health services and technologies generate huge volumes of data. The analysis of these data enables the transformation of conventional hypothesis-based information analysis into innovative data-driven analysis, capable of identifying links between heterogeneous information (Wang et al. 2016; Saranya et al. 2019).

1.6.3.2. *Artificial intelligence*

AI is attracting a great deal of interest due to its ability to process large quantities of data, produce accurate results and control processes to generate optimized outcomes. It is being used to aid decision-making and predict the effects of diseases, as well as longer term consequences (Kaur 2022). AI can perform processes such as logical reasoning, knowledge-based learning, drug discovery, guided surgery and advanced imagery (Sobhan et al. 2021).

1.6.3.3. *Robotics*

To ensure continuous, personalized care for patients in hospitals, or nursing homes, or homecare, solutions involving the use of robots are being proposed. These intelligent machines will help patients perform simple daily gestures, facilitate remote monitoring and communication with medical staff or relatives, administer simple therapies or be used for entertainment purposes (reading, storytelling, playing, etc.).

In addition to this type of robot, devices and control strategies for rehabilitation are being designed, such as the development of agents that can interact with the patient and provide real-time data to medical staff (Mashayekhi et al. 2020).

1.6.3.4. *Cloud*

It is known as a paradigm in which IT resources are made accessible to users. It offers many advantages, such as flexibility, cost and energy savings, resource sharing and rapid deployment. The rapid growth of e-health systems to deliver quality medical services has led to the use of cloud-based solutions. This choice makes it possible to take advantage of cloud resources to store and process large volumes of medical data.

1.6.3.5. *Internet of Things*

The IoT refers to a network of physical objects connected to the Internet and capable of communicating and exchanging data with each other and with other systems. These objects, also known as IoT devices, are equipped with sensors, software and communication technologies that enable them to collect, analyze and transmit information.

Currently, the approach used in most smart applications is to store all sensor data in the cloud and perform machine-learning processing on these data. The two worlds of IoT and cloud have seen rapid progress in the medical field. IoT can take advantage of the cloud's almost limitless resources to compensate for its insufficient capabilities. The main drivers of IoT integration in the cloud are as follows (Farahani et al. 2018; Yang et al. 2022):

– Communication: IoT is heterogeneous by nature and relies on a variety of communication protocols. The cloud offers an efficient solution for registering, discovering and managing any type of object, regardless of communication protocol.

– Resource pooling: physical IoT resources can be integrated into cloud resources, enabling us to allocate and share them on demand.

– Storage: IoT generates an enormous amount of data, characterized by its volume, variety and speed of data generation. In this context, IoT benefits from large-scale, long-term storage.

– Computing: data processing is generally a resource-intensive task. As a result, IoT can benefit from the cloud's unlimited processing resources to aggregate data and perform real-time analysis on the data collected.

1.7. Security in e-health systems

In this section, we present the required security services and legal requirements for e-health systems.

1.7.1. *Security services*

The main security services required for data in an e-health system are as follows:

– Confidentiality: medical data must be protected from disclosure to unauthorized parties during transmission, processing and/or storage (Siva Bharathi et al. 2019).

– Availability: having access to data and being able to dispose of it at any time is paramount in an e-health context. Given the critical applications of e-health, any disruption to the system can be life-threatening. Therefore, the ability to access the required data under any circumstances is essential.

– Integrity: integrity ensures that health data captured by a system or supplied to any entity is accurate and consistent with the information intended, without having been altered. Inappropriate treatment based on incorrect data can have serious consequences for patient health.

– Data authentication: unlike data integrity, which aims to prevent data being altered during transmission, data authentication aims to ensure that the message received comes from the originating node.

– Access control: it is a mechanism that limits access to legitimate entities. Access control policy is generally based on the privilege and right of each authorized entity. Several solutions have been proposed to address access control issues, among which role-based access control (RBAC) and attribute-based access control (ABAC) are the most popular models for e-health (Sandhu 1996).

– Data freshness: an attacker may intend to capture transmitted messages and replay them later, leading to confusion and instability in the network. It is therefore essential to have a mechanism in place to verify that the message received is fresh.

– Non-repudiation: repudiation threats concern users who deny having had access to medical data after consulting it.

1.7.2. *Legal environment for e-health systems*

Devices and applications that manage medical data must comply with regulations. In Europe, a regulation on the protection of personal data processing was adopted in 2018 (Koren et al. 2022). This directive is called the general data protection regulation (GDPR). It aims to harmonize data protection standards at European level and defines the rights and responsibilities of entities in charge of data processing. In the medical field, this directive specifies that personal health data are considered regardless of the source, for example, from a medical device (Koren et al. 2022). This directive imposes the concept of security by design, which means that security requirements must be considered right at the start of the solution design process.

In the United States, personal health data are governed by the Health Insurance Portability and Accountability Act (HIPAA) (Koren et al. 2022). HIPAA establishes security rules for all healthcare providers who process health information in electronic format. The law protects all individually identifiable health-related information held or transmitted by healthcare providers and which is held or transmitted by an entity. Indeed, under the rules defined by HIPAA, an entity (healthcare provider, health service provider, etc.) is responsible for the security of patient data, which means that an application or service that collects, processes and stores patient data must ensure data confidentiality and integrity and impose access restrictions.

1.8. Medical data security

In this section, we present the security techniques most widely used in the medical field. These include cryptography, biometrics and blockchain. A comparative analysis is carried out to identify the contribution of the techniques examined.

Figure 1.5 illustrates the security services required in e-health systems, as well as the different security techniques most commonly used to mitigate attacks in medical environments.

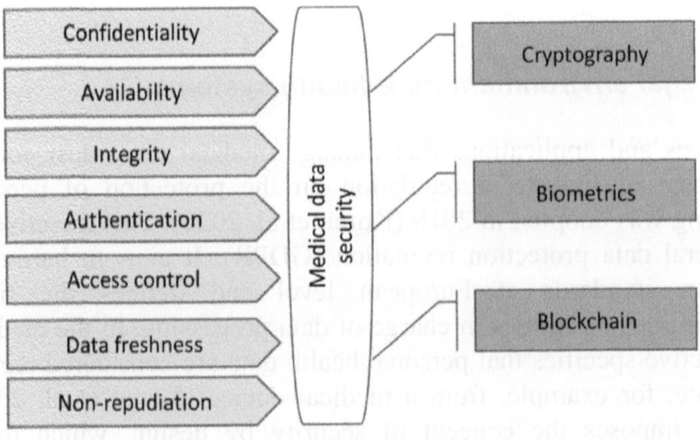

Figure 1.5. *Security in e-health systems*

1.8.1. *Cryptography*

Cryptography techniques offer important security services, such as confidentiality, integrity, non-repudiation and authentication of digital data (Anand et al. 2020; Singh et al. 2020).

Encryption is a technique used to encode information in order to guarantee the security of medical data (Thakur et al. 2018; Anand et al. 2020).

Some encryption techniques, such as public key cryptosystems or symmetric cryptosystems, have been developed to ensure confidentiality. In public-key cryptosystems, a public–private key pair is used to encrypt and decrypt the message. In the symmetric method, a single key is used to encrypt and decrypt the data (Qadir et al. 2019; Gupta et al. 2020).

Vinoth et al. (2017) proposed a cryptographic approach to sharing and storing medical records in the cloud. The patient reserves the right to grant each party access to his or her record by sharing a key. Sujatha et al. (2013) proposed an approach for fast and secure transmission of electrocardiogram (ECG) data in a WBAN. They adapted an encryption technique (set partitioning in hierarchical trees [SPIHT]).

Sun et al. (2011) proposed an approach that ensures medical data security, including confidentiality, access control, integrity, accountability. This system ensures high security by encrypting medical records using searchable symmetric encryption (SSE) and public key encryption with keyword search (PEKS) algorithms and cryptographic algorithms using identity-based cryptography (IBC). This system is highly effective in emergency situations.

References	Type of Data	Algorithm	Proposition
(Sujatha et al. 2013)	ECG	SPIHT	Secure ECG transmission using encryption, compression and genetic algorithms.
(Vinoth et al. 2017)	Medical records	Blowfish	Storing and sharing medical records in unsecured environments.
(Sun et al. 2011)	Medical records	SSE, PEKS, IBC	Protecting patient privacy in emergency situations.
(Nagamani et al. 2018)	Medical data	MD5, password	Authentication and security of medical data in the cloud.
(Sammoud et al. 2020)	Session key	ECG	Session key exchange in a WBAN.
(Sharma et al. 2021)	Health data	Quantum	Securing IoT-based healthcare systems.
(Jegadeesan et al. 2020)	Health data	Authentication	Anonymous authentication preserves patient privacy while reducing communication and computing costs.

Table 1.2. *Cryptography for e-health*

Nagamani et al. (2018) proposed a secure e-health system to predict patient condition based on problems and symptoms described by the user. This approach uses a system to study symptoms and provide an appropriate solution. To guarantee the confidentiality of transmitted information, it is encrypted using password-based encryption (PBE). The system also generates an encryption key from the user's password. In addition, a checksum is generated using the MD5 hash algorithm, and this value is appended to the transmitted data for user authentication.

Sammoud et al. (2020) proposed a session key exchange protocol using ECG. Sharma and Bhatt (2021) used quantum mechanics to secure IoT-based healthcare systems. Jegadeesan et al. (2020) presented an anonymous authentication scheme that preserves patient privacy while reducing communication and computational costs. Table 1.2 presents some cryptography-based security proposals.

1.8.2. *Biometrics*

Biometrics is the science of establishing an individual's identity based on physical or behavioral characteristics. The most common examples of biometric characteristics are DNA, fingerprint, iris, face, keystroke, smell, signature, retina, voice and hand veins. Biometric systems normally operate in two modes: verification and identification. In verification mode, the system validates the person's identity by matching the captured biometric information with its own biometric template(s) stored in the database. In identification mode, the system identifies a person by looking for a match between the templates of all the users in the database. As a result, the verification system performs a one-to-many comparison to establish an individual's identity. Biometrics has been used to improve the security and confidentiality of patient data in e-health applications.

In Jahan et al. (2019), an efficient and robust biometric system is proposed to authenticate the user and preserve the confidentiality of medical records. An approach using ECG-based biometrics is proposed for verifying patient identity and maintaining confidentiality of medical records. The specificity and availability of the ECG make it a suitable biometric choice.

The signal is captured, processed and then compared with the sample model; the result obtained indicates whether the user is authentic or not.

Ali et al. (2018) used biometric authentication. It operates in several phases: initialization, registration, login and authentication. To provide multi-level authentication, both the patient's fingerprints and the smart card are verified for login. In addition, mutual authentication and session key exchange are performed between the user and the medical server.

After studying a variety of attacks when accessing electronic medical records in a cloud environment, Hathaliya et al. (2019) developed an authentication and key exchange system based on fingerprint biometrics. The system also encrypts biometric data and a 160-bit random number and performs mutual authentication using a message authentication code (MAC). Sammoud et al. (2020) have proposed a multifactor authentication protocol using ECG biometric data. Table 1.3 presents some biometric-based security proposals.

References	Type of Data	Algorithm	Proposition
(Ali et al. 2018)	Medical data	Digital fingerprint + hash function	Secure data transmission between patient and healthcare professional.
(Jahan et al. 2019)	Base of ECG	ECG	User authentication to safeguard the security of medical records.
(Hathaliya et al. 2019)	Medical record	Digital fingerprint	Fingerprint authentication for secure access to medical records.
(Sammoud et al. 2020)	Medical data	ECG	Multi-factor authentication using ECG.

Table 1.3. *Biometrics for e-health*

1.8.3. *Blockchain*

Blockchain is one of the technologies used in e-health systems. It consists of a list of blocks in a distributed connected structure (Casino et al. 2018). It stores transactions and groups them in a structure (Li et al. 2020). Each block is stored at all network nodes in chronological order. Blockchain has solved the problem of centralization in a distributed system, bypassing single-point-of-failure issues and reducing transaction costs. Since no central authority is involved, transaction speed is also increased (Li et al. 2020).

Maintaining the integrity and confidentiality of medical records is paramount (Singh et al. 2017). The immutable property of blockchain ensures this (Agbo et al. 2019). In addition, it also guarantees data security and confidentiality. The decentralized structure of the blockchain can be implemented in healthcare systems, helping to preserve the confidentiality of sensitive data. Since blocks are replicated at each network node, the likelihood of data loss decreases and patients can also control access to their records (Agbo et al. 2019).

A system is proposed in Chen et al. (2018) for secure storage and communication using the blockchain of medical records between patients, medical institutes and third-party agencies. Patient details as well as medical records are encrypted using asymmetric encryption. What is more, the patient can authorize or withdraw access at any time.

An access control method is proposed in Fan et al. (2018), combining the concepts of blockchain and cryptography. It provides a solution for managing and sharing a large volume of medical data. Asymmetric encryption is used to secure medical records. The hash value of encrypted data is added to verify data integrity and authenticity.

Ichikawa et al. (2017) have designed an app for patients with insomnia, which allows timely scheduling of sessions based on patient-supplied data. JavaScript Object Notation (JSON) format is used to store the data. The proposed system also offers tamper-resistance using the Practical Byzantine Fault Tolerance (PBFT) consensus algorithm. Data can be successfully updated even in situations such as a network failure.

Garg et al. (2020) proposed a low-cost protocol with authentication in the IoT environment using blockchain. Islam and Young Shin proposed a blockchain-based healthcare system to help unmanned aerial vehicles (UAVs) secure health data collected from users and store it on the nearest server on the UAV's path (Islam et al. 2020). Miyachi et al. (2021) presented a blockchain-based system that preserves patient privacy.

Table 1.4 presents security proposals based on the blockchain.

References	Type of Data	Algorithm	Proposition
(Ichikawa et al. 2017)	Insomnia patient data	PBFT	Non-violation of medical data when updating.
(Chen et al. 2018)	Indexing information for medical data and transaction records	Encryption	Privacy protection, secure sharing and storage of medical data.
(Fan et al. 2018)	Medical information	Asymmetric encryption and hashing	Secure data sharing and controlled access to medical data.
(Garg et al. 2020)	Medical information	Hashing	Access control.
(Islam et al. 2020)	Patient data in UAVs	Hashing	Securing health data collected from users in the UAV and storing it on the server closest to the UAV.
(Miyachi et al. 2021)	Confidential patient data	Encryption and hashing	Protecting patient privacy.

Table 1.4. *Using the blockchain in e-health*

1.9. Perspectives

– Data sharing: data generated by different devices can pose a variety of security and confidentiality problems for e-health systems, as patient data are sensitive and should not be shared by all medical staff. It is therefore necessary to adapt security countermeasures to guarantee the security, confidentiality and access control of medical data and devices.

– Medical data: a considerable amount of data is collected from various devices. These data are constantly changing as a function of the patient's state of health. Managing this amount of information is a time-consuming task. What is more, different types of data need to be stored in different formats. It is therefore necessary to have an efficient data management system capable of converting the file according to the needs of each medical application.

– Lack of standardization: many devices are used in e-health systems to measure, collect and relay health data. Each device has its own set of protocols and configurations for sharing this information with medical staff. However, there is no centralized consensus or standardization available for the communication, implementation and deployment of these devices. It is therefore necessary to research this aspect, so that devices using different standards and protocols can communicate.

1.10. Conclusion

In the digital age, e-health is playing an increasingly important role in the management of chronic diseases and in communication between healthcare professionals and patients. Thanks to technological advances, it is now possible to ensure continuous, real-time disease management, both inside and outside medical structures.

The aim of e-health is to improve healthcare services using digital technologies. This makes healthcare more accessible, efficient and personalized for a greater number of people, contributing to better health and well-being.

In this context, various e-health architectures have been developed to integrate digital technologies into healthcare systems. A simplified architecture has been presented, bringing together the different technologies used in this field. This architecture makes it possible to visualize the various components and interactions between healthcare system actors, connected medical devices, mobile applications, data platforms and so on.

The security of e-health systems is also an essential aspect examined in this chapter. Healthcare data are sensitive and confidential, and it is crucial to implement robust security measures to protect them against cyber-attacks and unauthorized access. Various security techniques are used, such as user authentication and authorization, data encryption, access management and suspicious activity monitoring.

In short, e-health uses technological advances to improve healthcare services and promote continuous, real-time management of chronic diseases. e-Health architectures provide a global vision of the various technologies used, while security techniques guarantee data protection and the confidentiality of medical information. These advances contribute to improved accessibility and more effective, personalized healthcare for patients.

1.11. References

Agbo, C.C., Mahmoud, Q.H., Eklund, J.M. (2019). Blockchain technology in healthcare: A systematic review. *Healthcare*, 7(2), 56.

Ali, R. and Pal, A.K. (2018). Cryptanalysis and biometric-based enhancement of a remote user authentication scheme for e-healthcare system. *Arabian Journal for Science and Engineering*, 43, 7837–7852.

Anand, A. and Singh, A.K. (2020). Joint watermarking-encryption-ECC for patient record security in wavelet domain. *IEEE MultiMedia*, 27(3), 66–75.

Atienza, A.A., Hesse, B.W., Baker, T.B., Abrams, D.B., Rimer, B.K., Croyle, R.T., Volckmann, L.N. (2007). Critical issues in e-health research. *American Journal of Preventive Medicine*, 32(5), S71–S74.

Balakrishnan, L. (2021). An Internet of Things (IoT) based intelligent framework for healthcare – A survey. In *2021 3rd International Conference on Signal Processing and Communication (ICPSC)*. IEEE.

Basaklar, T., Tuncel, Y., An, S., Ogras, U. (2021). Wearable devices and low-power design for smart health applications: Challenges and opportunities. In *2021 IEEE/ACM International Symposium on Low Power Electronics and Design (ISLPED)*. IEEE.

Bashshur, R., Shannon, G., Krupinski, E., Grigsby, J. (2011). The taxonomy of telemedicine. *Telemedicine and e-Health*, 17(6), 484–494.

Behera, S.K. (2021). Chipless RFID sensors for wearable applications: A review. *IEEE Sensors Journal*, 22(2), 1105–1120.

Casino, F., Dasaklis, T.K., Patsakis, C. (2019). A systematic literature review of block-chain-based applications: Current status, classification and open issues. *Telematics and Informatics*, 36, 55–81.

Chen, Y., Ding, S., Xu, Z., Zheng, H., Yang, S. (2019). Blockchain-based medical records secure storage and medical service framework. *Journal of Medical Systems*, 43, 1–9.

Chung, Y.F. and Liu, C.H. (2013). Design of a wireless sensor network platform for tele-homecare. *Sensors*, 13(12), 17156–17175.

Devedžić, G., Koceski, S., Savić, S.P. (2021). A brief overview of enabling technologies for digital medicine and smart healthcare. In *2021 10th Mediterranean Conference on Embedded Computing (MECO)*. IEEE.

Fagroud, F., Ben Lahmar, E., El Filali, S. (2019). Internet of things: Statistical study on research evolution. *International Journal of Advances in Electronics and Computer Science*, 6(5), 4–13.

Fan, K., Wang, S., Ren, Y., Li, H., Yang, Y. (2018). Medblock: Efficient and secure medical data sharing via blockchain. *Journal of Medical Systems*, 42, 1–11.

Farahani, B., Firouzi, F., Chang, V., Badaroglu, M., Constant, N., Mankodiya, K. (2018). Towards fog-driven IoT eHealth: Promises and challenges of IoT in medicine and healthcare. *Future Generation Computer Systems*, 78, 659–676.

Garg, N., Wazid, M., Das, A.K., Singh, D.P., Rodrigues, J.J., Park, Y. (2020). BAKMP-IoMT: Design of blockchain enabled authenticated key management protocol for internet of medical things deployment. *IEEE Access*, 8, 95956–95977.

Gourob, J.H., Raxit, S., Hasan, A. (2021). A robotic hand: Controlled with vision based hand gesture recognition system. In *2021 International Conference on Automation, Control and Mechatronics for Industry 4.0 (ACMI)*. IEEE.

Gupta, B.B., Perez, G.M., Agrawal, D.P., Gupta, D. (2020). *Handbook of Computer Networks and Cyber Security*. Springer, Cham.

Gupta, M., Tanwar, S., Rana, A., Walia, H. (2021). Smart healthcare monitoring system using wireless body area network. In *2021 9th International Conference on Reliability, Infocom Technologies and Optimization (Trends and Future Directions) (ICRITO)*. IEEE.

Hajar, M.S., Al-Kadri, M.O., Kalutarage, H.K. (2021). A survey on wireless body area networks: Architecture, security challenges and research opportunities. *Computers & Security*, 104, 102211.

Hamdi, O., Chalouf, M.A., Ouattara, D., Krief, F. (2014). eHealth: Survey on research projects, comparative study of telemonitoring architectures and main issues. *Journal of Network and Computer Applications*, 46, 100–112.

Hathaliya, J.J., Tanwar, S., Tyagi, S., Kumar, N. (2019). Securing electronics healthcare records in healthcare 4.0: A biometric-based approach. *Computers & Electrical Engineering*, 76, 398–410.

Hermes, S., Riasanow, T., Clemons, E.K., Böhm, M., Krcmar, H. (2020). The digital transformation of the healthcare industry: Exploring the rise of emerging platform eco-systems and their influence on the role of patients. *Business Research*, 13, 1033–1069.

Hodgkiss, J. and Djahel, S. (2020). Securing fuzzy vault enabled authentication in body area networks-based smart healthcare. *IEEE Consumer Electronics Magazine*, 11(1), 6–16.

Ichikawa, D., Kashiyama, M., Ueno, T. (2017). Tamper-resistant mobile health using blockchain technology. *JMIR mHealth and uHealth*, 5(7), e7938.

Islam, A. and Shin, S.Y. (2020). A blockchain-based secure healthcare scheme with the assistance of unmanned aerial vehicle in Internet of Things. *Computers & Electrical Engineering*, 84, 106627.

Jahan, S., Chowdhury, M., Islam, R. (2019). Robust user authentication model for securing electronic healthcare system using fingerprint biometrics. *International Journal of Computers and Applications*, 41(3), 233–242.

Javaid, S., Zeadally, S., Fahim, H., He, B. (2022). Medical sensors and their integration in wireless body area networks for pervasive healthcare delivery: A review. *IEEE Sensors Journal*, 22(5), 3860–3877.

Jegadeesan, S., Azees, M., Babu, N.R., Subramaniam, U., Almakhles, J.D. (2020). EPAW: Efficient privacy preserving anonymous mutual authentication scheme for wireless body area networks (WBANs). *IEEE Access*, 8, 48576–48586.

Kaur, D., Uslu, S., Rittichier, K.J., Durresi, A. (2022). Trustworthy artificial intelligence: A review. *ACM Computing Surveys (CSUR)*, 55(2), 1–38.

Koren, A. and Prasad, R. (2022). IoT health data in electronic health records (EHR): Security and privacy issues in era of 6G. *Journal of ICT Standardization*, 10(1), 63–84.

Kostkova, P. (2015). Grand challenges in digital health. *Frontiers in Public Health*, 3, 134.

Légifrance (2010). Décret No. 2010-1229 du 19 octobre 2010 relatif à la télémédecine [Online]. Available at: https://www.legifrance.gouv.fr/jorf/id/JORFTEXT000022932449/ [Accessed 22 February 2023].

Li, X., Jiang, P., Chen, T., Luo, X., Wen, Q. (2020). A survey on the security of blockchain systems. *Future Generation Computer Systems*, 107, 841–853.

Mashayekhi, A., Behbahani, S., Ficuciello, F., Siciliano, B. (2020). Influence of human operator on stability of haptic rendering: A closed-form equation. *International Journal of Intelligent Robotics and Applications*, 4, 403–415.

Minakshi, J.A.H. (2016). An overview of wireless body area network (WBAN) using Zigbee technology. *International Journal of Scientific Development and Research (IJSDR)*, 1, 2455–2631.

Miyachi, K. and Mackey, T.K. (2021). hOCBS: A privacy-preserving blockchain framework for healthcare data leveraging an on-chain and off-chain system design. *Information Processing & Management*, 58(3), 102535.

Nagamani, S. and Nagaraju, D.R. (2018). A mobile cloud-based approach for secure m-health prediction application. *International Journal for Innovative Engineering & Management Research*, 7(12).

Negra, R., Jemili, I., Belghith, A. (2016). Wireless body area networks: Applications and technologies. *Procedia Computer Science*, 83, 1274–1281.

Qadir, A.M. and Varol, N. (2019). A review paper on cryptography. In *2019 7th International Symposium on Digital Forensics and Security (ISDFS)*. IEEE.

Rabbani, R., Najafiaghdam, H., Ghanbari, M.M., Papageorgiou, E.P., Zhao, B., Roschelle, M., Stojanovic, V., Muller, R., Anwar, M. (2021). Towards an implantable fluorescence image sensor for real-time monitoring of immune response in cancer therapy. In *2021 43rd Annual International Conference of the IEEE Engineering in Medicine & Biology Society (EMBC)*.

Sammoud, A., Chalouf, M.A., Hamdi, O., Montavont, N., Bouallegue, A. (2020). A new biometrics-based key establishment protocol in WBAN: Energy efficiency and security robustness analysis. *Computers & Security*, 96, 101838.

Sandhu, R.S., Coyne, E.J., Feinstein, H.L., Youman, C.E. (1996). Role-based access control models. *Computer*, 29(2), 38–47.

Saranya, P. and Asha, P. (2019). Survey on big data analytics in health care. In *2019 International Conference on Smart Systems and Inventive Technology (ICSSIT)*. IEEE.

Sharma, A. and Bhatt, A.P. (2021). Quantum cryptography for securing IoT-based healthcare systems. In *Limitations and Future Applications of Quantum Cryptography*, Kumar, N., Agrawal, A., Chaurasia, B.K. (ed.). IGI Global, Hershey, PA, 124–147.

Singh, A.K. and Kumar, C. (2020). Encryption-then-compression-based copyright protection scheme for E-governance. *IT Professional*, 22(2), 45–52.

Singh, A.K., Kumar, B., Singh, G., Mohan, A. (2017). *Medical Image Watermarking*. Springer Science Business Media, Berlin.

Siva Bharathi, K.R. and Venkateswari, R. (2019). Security challenges and solutions for wireless body area networks. In *Computing, Communication and Signal Processing: Proceedings of ICCASP 2018*. Springer, Berlin.

Sobhan, S., Islam, S., Valero, M., Shahriar, H., Ahamed, S.I. (2021). Data analysis methods for health monitoring sensors: a survey. In *2021 IEEE 45th Annual Computers, Software, and Applications Conference (COMPSAC)*.

Sujatha, S. and Govindaraju, R. (2013). A secure crypto based ECG data communication using modified SPHIT and modified quasigroup encryption. *International Journal of Computer Applications*, 78(6).

Sun, J., Zhu, X., Zhang, C., Fang, Y. (2011). HCPP: Cryptography based secure EHR system for patient privacy and emergency healthcare. In *2011 31st International Conference on Distributed Computing Systems*. IEEE.

Sundaram, J.P.S., Wan, D., Zhao, Z. (2019). A survey on LoRa networking: Research problems, current solutions and open issues. *IEEE Communications Surveys & Tutorials*, 22, 371–388.

Thakur, S., Singh, A.K., Ghrera, S.P., Dave, M. (2018). Watermarking techniques and its applications in tele-health: A technical survey. In *Cryptographic and Information Security: Approaches for Images and Videos*, Ramakrishnan, S. (ed.). CRC Press, New York.

Wang, K., Shao, Y., Shu, L., Zhu, C., Zhang, Y. (2016). Mobile big data fault-tolerant processing for e-health networks. *IEEE Network*, 30(1), 36–42.

Yang, Y., Wang, H., Jiang, R., Guo, X., Cheng, J., Chen, Y. (2022). A review of IoT-enabled mobile healthcare: Technologies, challenges, and future trends. *IEEE Internet of Things Journal*, 9(12), 9478–9502.

2

Vulnerabilities in e-Health and Countermeasures

Aida BEN CHEHIDA DOUSS and Ryma ABASSI
Innov'COM, SUP'COM, Université de Carthage,
Tunis, Tunisia

2.1. Introduction

Every year, the number of cyber-attacks is increasing. All sectors of activity are affected by these attacks: private companies, institutions, schools, local authorities, government bodies and so on. Healthcare structures, with their digital boom and massive use of technology, have also become highly prized by attackers. Hospitals and laboratories are the main targets. Several factors may explain this apparent upsurge in attacks against the healthcare sector. First and foremost, there is the massive computerization of care procedures and digital services, the use of connected equipment with complex IT systems and the management of large quantities of precious and sensitive data concerning both patients and caregivers. What is more, healthcare facilities have an imperative need for uninterrupted services.

All these parameters make this sector an attractive target for cybercriminals with different profiles and motivations. Indeed, several attacker profiles exist. We can mention white hat hackers, who are the computer experts behind the main protocols we use; black hat hackers, who are hackers whose aim is to harm the system under attack; and gray hat hackers, who attack systems, but then warn those responsible of the security flaws they have found. In this chapter, we will be looking at hackers who may have a variety of motivations for breaking in and attacking a particular hospital system. Their goals can be as varied as stealing sensitive patient and caregiver data, deleting it, making it public or selling it on the black market. Attackers can also block the normal operation of hospital services. All these attacks can disrupt the proper functioning of e-health systems, and can cause a variety of damage, even endangering patients' lives.

In order to secure hospital structures, several security mechanisms and measures have been proposed in the literature. These measures are mainly based on cloud computing and the Internet of Things (IoT) and can be divided into a set of layers providing certain services and using the services of the previous layers to function properly.

It is within this framework that our chapter is situated, with the main aim of looking at the e-health sector, and more specifically at the cyber-attacks that can perpetrate it, and the mechanisms and security measures that can protect it. To this end, this chapter is divided into two main sections. In the first part, we first present the importance and challenges of digitization in the healthcare sector. We then go on to describe the profiles and motivations of cybercriminals. Next, we shall look at the risks and consequences of attacks on the healthcare system, and the different types of attack that can be carried out by attackers. In the second part of the chapter, we will review some of the main security incidents against e-health systems that have been reported in the literature in recent years. We will then outline security measures for e-health systems based on a layered architecture. Finally, we will present a number of mechanisms and recommendations that can be put in place to alleviate any security problems that may affect e-health systems. These mechanisms can be classified into risk management methods, in-depth defense architectures and awareness-raising and training initiatives. This chapter will then close with a conclusion and overview of prospective futures.

2.2. The importance of digitization in healthcare systems

Healthcare services are in the throes of technological change. Indeed, thanks to digitization, the healthcare sector has experienced a major technological boom, enabling professionals to better coordinate and improve patient care. Digitization offers enormous potential for healthcare systems, not only to enhance patient well-being and safety, but also to promote disease prevention and early detection (Patru 2022).

The last two years, marked by the Covid-19 pandemic, have clearly demonstrated the importance of digitization in the healthcare sector. Indeed, sharing and exchanging information in real time on contamination test results, genome sequencing of virus variants and hospital bed occupancy rates has been of vital importance. This enabled the authorities concerned to take a global view of the pandemic, so that they could quickly take the best decisions to curb the circulation of the virus and give patients greater autonomy in screening and monitoring their symptoms. The Covid crisis has also contributed to the emergence of telemedicine. Doctors can now provide remote medical follow-up for their sick patients, without losing touch with them. Thanks to teleconsultation, patients in remote areas suffering from a lack of specialist doctors can have access to care. In this way, the risk of disease complications linked to irregular follow-up is reduced by continuous monitoring, even at a distance.

Another key area for digitization in the healthcare sector concerns medical records. In fact, digitizing patient medical records means doing away with paper files, which has proved ineffective, given the time wasted in filing them, the large amount of space they take up in archives, the effort wasted, not to mention their cost and loss over time. A digital medical records management platform thus makes it possible to move toward a more fluid care system, and above all to simplify the patient's hospital journey (from the detection of symptoms, to direct or remote care, right through to recovery).

Digitization in the healthcare sector has also demonstrated its importance in the field of drug and medical supplies inventory management. Indeed, this digitization process enables hospital establishments to effectively control their stock movements and requirements, with a view to avoiding drug shortages, but also to control their expenditure, with a view to adopting a forward-looking vision in terms of managing hospital budgets.

All these aspects of digitization mentioned in this section have demonstrated the importance of technology in the healthcare sector.

2.3. The challenges of digitization in e-health systems

The digital transformation of traditional healthcare processes is having a major positive impact, as explained in section 2.2. Indeed, this digitization enables easier access, higher quality and more relevant care, faster patient care, simpler and stronger doctor–patient relationships, better management of medical records and drug stocks, and more effective management of pandemics. As a result, the digitization of the healthcare sector brings considerable benefits to professionals, patients and, more generally, all citizens (Neumann et al. 2021).

Nevertheless, the digitization of healthcare services and the introduction of cloud-based infrastructures (for the storage and management of huge amounts of data handled by medical facilities) are increasing the presence of vulnerabilities within healthcare establishments and posing a key problem: where there is technology, there is a need for security (Bhambere et al. 2021). However, the technological structure of healthcare organizations is often obsolete and unsuitable in terms of security standards, making them an Eldorado for hackers. Hackers spend most of their time trying to infiltrate e-health systems, carrying out attacks that are increasingly complex and difficult to detect. As a result, a cyber-attack cannot only disrupt the daily lives of healthcare professionals, but also jeopardize the care and lives of patients. By way of example, we could mention connected medical devices, which security professionals need to pay particular attention to. The question obviously arises as to what would happen if, tomorrow, a surgical robot came under the control of an attacker. Or if a cyber-attack were to hijack the continuity of service in operating theatres or take control of an entire hospital or its electrical network or falsify vital patient data. All this could have far-reaching consequences for the smooth running of the hospital and put patients' lives at risk.

Faced with these risks, it has now become essential to integrate security into the e-health services process, in order to take the appropriate measures to guard against the risk of attack, and to avoid as far as possible the collapse of this sector, which is so important at every level. As a result, the healthcare sector, like industry, needs specific cybersecurity solutions to ensure protection, data confidentiality and continuity of services.

2.4. Cyber-attacks in the healthcare sector

Technological advances in the healthcare sector have led to increased interconnectivity and ever-greater cloud infrastructures, making it possible to manage a varied flow of data (i.e. medical records, medication stocks, etc.) and to treat large numbers of patients via telehealth. Nevertheless, as mentioned above, the healthcare sector is often a relatively easy target for hackers. Many elements make e-health systems vulnerable; we can cite the rise of the Internet of Medical Things (IoMT), insufficient protection of portals for sharing patients' medical information (which can be used as an entry point via weak passwords), the non-use of multi-factor authentication (MFA), the lack of cybersecurity solutions capable of stopping sophisticated threats, but also ineffective employee training. All these vulnerabilities make attacks on healthcare systems easier to perpetrate.

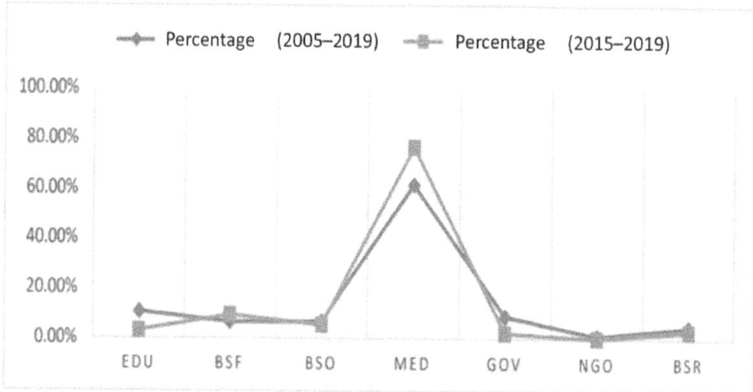

Figure 2.1. *Representation of data breach incidents by industry sector (Seh et al. 2020)*

A study by Adil Hussain Seh et al. (2020) analyzed data breach rates by sector over the period from 2005 to 2019. Various sectors were considered in this study, namely the education sector (EDU), financial companies (BSF), trading companies (BSR), other companies (BSO), healthcare service providers (MED), government and defense institutes (GOV) and non-governmental organizations (NGOs). The in-depth analysis of the entire 15-year period, carried out in this study, showed that the healthcare sector (MED), in the two periods of (2005–2019) and (2015–2019), faced the highest number of data breaches. Indeed, of the 6,355 breach incidents reported between 2005 and 2019, 3,912 were recorded in the healthcare sector. This represents 61.55% of total attacks. The medical sector is followed by the education and government sectors, accounting for 10.55% and 8.82%, respectively. Over the last 5 years (2015–2019), 2,072 attacks have been recorded, 1,587 of which represent attacks on the healthcare system (76.59%). Figure 2.1, presented by Seh et al. (2020), shows a graphical representation of the analysis that was carried out, and concludes that all sectors experienced a decrease in the percentage of attacks in the five years from 2015 to 2019, except for the medical sector, which was marked by an increase in the data breach rate. The graph therefore indicates that the healthcare industry is a prime target for attackers.

Before we look at the main attacks on e-health systems, we will start by describing the profiles of cybercriminals, and then go on to look at the different motivations that drive attackers to carry out attacks on e-health systems.

2.4.1. *Profiles of cybercriminals*

Cybercriminals, or hackers, have evolved over time. In fact, there are many different types of hackers today. We can list three main categories: white hat hackers, black hat hackers and gray hat hackers (Kavallieros et al. 2021).

White hat hackers are computer security experts. They carry out penetration tests and other testing methods to ensure the security of information systems. Their main aim is to help improve computer systems and technologies, and they are generally behind the main computer protocols and tools we use today. This type of hacker is also known as an ethical hacker, Pentester (for penetration test), or benevolent hacker. Their slogan is "learn to attack to better defend yourself" (and do not aim to cause any damage or disruption).

Black hat hackers, on the other hand, are cybercriminals with malevolent intent. Their aim is to exploit their skills and knowledge for the wrong reasons. They deliberately damage corporate IT systems through a variety of methods, including scams, data theft and espionage. Black hat hackers are also simply known as hackers.

Gray hat hackers are a cross between the two other types. These hackers sometimes act for a good cause, as a white hat hacker would, but may occasionally commit crimes, like a black hat hacker. They may, for example, illegally break into a system, only to warn those responsible of the flaws they have found. Unlike ethical hackers, who have permission to test system vulnerabilities, gray hat hackers dispense with this permission and infiltrate systems illegally. Their actions are commendable, but illegal all the same.

However, other hacker profiles also exist. We could mention script-kiddies, who, lacking basic computer security skills, spend most of their time trying to infiltrate systems, using scripts or programs developed by others. These hackers are generally young and have little knowledge. We can also mention *hacktivists*, who are hackers with a particular objective: they target politics to undermine the systems of governments, global organizations and large multinational corporations. Their main aim is to bring about political or societal change.

In what follows, we will focus on black hat hackers. Throughout this chapter, they will be referred to as cybercriminals or hackers.

2.4.2. *Motivations of cybercriminals*

Healthcare establishments control a huge amount of data. These data are many and varied. They may relate to a patient's personal information collected at the time of registration (surname, first name, address, social security number, etc.), to information concerning a disease or to information obtained during the testing or examination of one or more parts of a patient's body (Chng et al. 2022). Access to these data and the proper functioning of hospital IT systems are essential. If access to the IT system, and therefore to the data, is paralyzed, the entire hospital is in a very difficult position to carry out its mission, and the consequences can be very serious, both for patients and for the e-health system in general. As a result, cybercriminals are increasingly motivated when attacking medical establishments. The three main motivations we will be looking at, and which drive the vast majority of cybercriminals, are money, information espionage and the desire to cause harm. According to a Verizon Enterprise report (Data Breach Investigations 2022), financial and information espionage motivations account for 93% of attack motives (Sütterlin et al. 2022).

2.4.2.1. *Money*

Cybercriminals' primary motivation is purely financial. Whether through ransomware (viruses blocking access to data) or data theft, cybercriminals' demands are always the same: Access to stolen or blocked data will be restored to you in exchange for a ransom payment. Hackers can also encrypt the data of healthcare institutions and decrypt it for a fee. This is data hacking for the sake of hacking.

Faced with such situations, many healthcare facilities are tempted and obliged to pay the ransoms demanded, given the vital nature of their data. This can lead to very serious situations for hospitals. Indeed, since hospitals already lack the means to fulfill their primary missions of patient care and support (their IT systems suffer from a lack of funds and protection), a cyber-attack involving the payment of ransoms can only aggravate the situation. What is more, these establishments know that every second can be vital for a patient. Cybercriminals are aware of this situation and know that it will put additional pressure on hospitals to pay ransoms.

2.4.2.2. *Information*

Another of cybercriminals' main motivations is information, and more specifically the resale of patient health information to supplementary insurance companies and pharmaceutical laboratories. Indeed, medical information can prove invaluable to insurers, enabling them to adjust rates more precisely. As for the pharmaceutical industry, this information can be used to compile vital statistics, with the aim of orienting drug marketing. A pharmaceutical company could use these statistics to direct drug sales in a given territory, according to the specific problems of its population. Another reason for attackers to use information is industrial espionage. Hackers can steal data relating to research or medical advances from pharmaceutical laboratories in order to resell them to competitors.

2.4.2.3. *Other motivations of cybercriminals*

Other, less frequent motivations for cybercriminals exist, and relate mainly to pleasure-seeking, ideology, the desire to cause harm, etc. Indeed, some cybercriminals take pleasure in exploiting the weaknesses of e-health IT systems for ideological reasons, or simply for the pleasure of divulging information. Finally, there is a population of cybercriminals who, simply out of spite (because they are dissatisfied with a service, for example), may use any of these attacks to satisfy their personal motivations.

2.4.3. *Risks and repercussions*

Attacks on the healthcare sector are far from risk-free and without consequences. Indeed, following an attack, all vital systems are compromised: from the information system to the communications system, but also the medical equipment used, such as scanners, magnetic resonance imagings (MRIs), infusion pumps and so on. As a result, the entire system can be paralyzed in a matter of seconds, and patients' lives can be put at real risk. The risks and consequences of an attack on a healthcare facility are as follows:

– Risk of manual file processing and a return to paper and pencil: An attack targeting a hospital's computer network or blocking access to its various departments can lead to caregivers manually processing patient files to manage admissions, prescriptions and reports. This can contribute considerably to slowing down patient care and disorganizing consultation and surgery schedules.

– Risk unavailability of services: An attack on the hospital system may result in the interruption or unavailability of its services. An attack of this kind can have very serious consequences if, for example, it targets an online tool essential to the smooth running of the hospital. Even a brief interruption to a given service can have disastrous repercussions on patient follow-up and, above all, on patient care.

– Risk of disclosure of sensitive personal data: An attack on an e-health system can lead to the disclosure of confidential patient and caregiver data. Hackers can then use these personal data for identity theft, black-market sales or blackmail. In the third case, the hacker generally demands a ransom, usually a large sum of money, in exchange for unblocking the network or withholding personal information.

– Risk of additional costs for e-health systems: As outlined above, an attacker can steal medical data in order to blackmail victims into paying ransom. This represents an additional cost for hospitals already suffering from a lack of funds. In addition, penalties and sanctions can be imposed by commissions such as the CNIL[1] for failure to respect the confidentiality of sensitive health data.

– Risk of data modification: attacks on e-health systems can also result in modifications to patients' personal and medical data, as well as to the diagnostic results of medical devices used by caregivers. This can have serious consequences for patient diagnosis and treatment.

2.4.4. *Types of attacks*

In the context of e-health systems, we consider a security attack to be a deliberate action that compromises the security and confidentiality of information (Zeadally et al. 2016). In what follows, we will present the cyber-attacks that can be carried out on the healthcare sector.

2.4.4.1. *Phishing*

A phishing attack is a scamming technique used by hackers to obtain personal or professional information by pretending to be a trusted third party. It is a social engineering technique that consists of fooling users by usurping the visual identity of a company or organization, with the aim of inciting them to communicate their personal data (access accounts, passwords, etc.).

In the health sector, where urgency is paramount, these attacks are particularly effective. Indeed, their aim is to deceive users (nurses, doctors, etc.), who are increasingly in a hurry and less attentive, in order to extract supposedly confidential information, money or access to services that are not normally authorized.

2.4.4.2. *Ransomwares*

Ransomware is malicious code that disables the victim's computer or information system, blocking access to personal data by encrypting it. The attacker then sends a message to the victim, offering to decrypt the data in return for payment of a ransom. To increase the pressure, the attacker may threaten to make the data public. Hackers usually demand payment of ransoms in cryptocurrencies to preserve their anonymity.

---

1 CNIL refers to France's *Commission Nationale de l'Informatique et des Libertés* (National Commission for Data Protection and Liberties).

During a ransomware attack on healthcare facilities, all vital systems can be compromised. The attack can bring information and communication systems, as well as medical equipment, to a standstill. This can lead to the paralysis of the entire e-health system in a matter of seconds and can be life-threatening for patients. The unavailability of health data can lead to delays in diagnosis and patient care, with potentially serious consequences for the patient, including death. As a result, without access to data and diagnostics, care providers will have to practice combat medicine.

2.4.4.3. *Data theft*

Medical data are very valuable. Indeed, it is characterized by a wealth of information that attracts the interest of attackers. Healthcare organizations manage a large volume of confidential data, the security of which is of paramount importance. These include patient data (surname, first name, telephone number, address, social security number, name of attending physician, patient's condition and pathologies, blood group, etc.), research data (statistics on diseases, drugs used, etc.) and the various login credentials used to infiltrate computer systems and perpetrate more complex attacks. This confidential information can be stolen and quickly sold on the black market to research laboratories, insurance companies (industrial espionage) or any other person with an interest in possessing this information (e.g. to create false identities).

2.4.4.4. *Attacks on medical devices*

Medical devices have revolutionized the healthcare sector. They have optimized the efficiency of diagnosis and patient follow-up because of automation. At the same time, they have considerably improved patients' state of health. To treat their patients, doctors and nurses use several types of equipment adapted to their needs. These can be categorized as follows: diagnostic equipment (CT scanners, MRI, etc.), laboratory equipment (blood and urine analyzers, etc.), life-support equipment (medical ventilators, extracorporeal membrane oxygenation, dialysis machines, etc.), which help to maintain the patient's bodily functions, plus monitoring devices (electrocardiography, electroencephalography, etc.), which periodically monitor and measure the patient's medical condition. All these medical devices are used in healthcare facilities on an ongoing basis. Indeed, many of them have 24/7 operating requirements, and a simple stoppage for maintenance, patching or updating, can significantly affect the work of caregivers and the health status of patients. What is more, once a piece of medical equipment has been assigned to a patient, they may use it for several days, or even weeks.

Given their vital importance and the medical records they may contain, these healthcare devices have become a target for attackers. An attack on medical devices, also known as a medjack (medical device hijack) attack, involves creating a backdoor on the target device, in order to take control of it. For example, if an attacker succeeds in taking control of an infusion pump, they will be able to administer inappropriate doses of medication to the patient, which could prove fatal. An attacker could also maliciously control a pacemaker, causing the patient's death, or decide to change the reading of an electrocardiogram, misleading the staff in charge.

Medical imaging has also become a target for attackers. Attackers can embed malware in MRI images, for example, modifying them (leading to misreading by treating physicians, diagnostic errors and a risk to patients' lives) or simply deleting them (leading to delayed diagnosis and an impact on data confidentiality). A concrete example of this vulnerability was discovered by security researchers in two models of hospital anesthesia machines manufactured by General Electric. According to the researchers, the vulnerabilities lie in the firmware of the two devices, which would allow attackers on a hospital network, for example, to send remote commands that could modify the devices' parameters (altering the concentration of oxygen, CO_2, N_2O and other anesthetic agents, or the barometric pressure of the gas), and harm diagnosis and patient life.

2.4.4.5. *DDoS attacks*

DoS (denial of service) attacks target service availability and aim to saturate a given service by sending massive numbers of requests until it reaches saturation point. Attacks of this type are just as devastating as ransomware attacks. Indeed, stopping a service, even for a limited time, can cause damage to medical structures. DoS attacks, or DDoS (distributed DoS) attacks, are generally targeted attacks aimed at making a given service unavailable but can be a preliminary attack that may conceal a more complex and elaborate attack. That is why, after a DoS attack, we must always remain vigilant and on the lookout for other possible attacks.

2.4.4.6. *Internal attacks*

Internal threats caused by dishonest employees are a real danger for healthcare organizations, given the trust placed in them. Indeed, a dishonest employee can abuse their access authorizations and the trust placed in them by the organization to perpetrate an attack. Internal threats can also emanate from unskilled employees (misconfigurations, poor installations, etc.) who may unwittingly cause major damage to healthcare facilities. One of the best ways to protect against insider attacks is to raise awareness of the importance of good cybersecurity practices among all healthcare professionals, right through the organization's hierarchy.

2.5. Security incidents in the healthcare sector

In this section, we review some of the main security incidents against e-health systems that have been reported in the literature in recent years (Agence France-Presse 2022).

Before outlining these incidents, we examine an online monitoring system set up by the World Health Organization (WHO) to record acts and threats of violence, as well as obstacles that compromise the availability or provision of preventive and curative health care and services in countries affected by conflict or crisis (OMS 2022). This tool targets several audiences. These include, of course, healthcare personnel and those responsible for security in healthcare structures, but also national and international governmental and non-governmental organizations, academia and, in general, anyone with an interest in the issue.

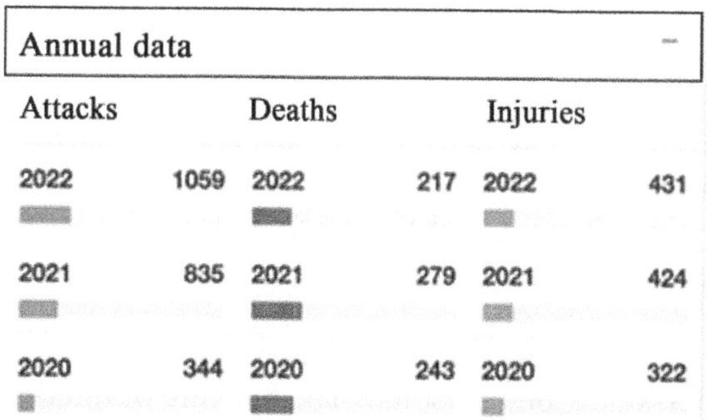

Figure 2.2. *Attack statistics for e-health 2020, 2021 and 2022 (WHO 2022)*

Figure 2.2 shows the results of attack statistics on e-health systems developed by an online tool, over the 3 years: 2020, 2021 and 2022. This figure shows that the number of attacks has risen dramatically, from 344 attacks in 2020 to 1,059 attacks in 2022. Thanks to the data collected, we will be able to gain a better understanding of global trends and patterns in this type of violence, as well as a clearer picture of the nature of the attacks and their impact on healthcare provision. The WHO is using this information as part of its ongoing awareness-raising campaign to stop attacks on healthcare, and to raise awareness of the relevant rules of international humanitarian law. The ultimate aim is to encourage the international community to take action to better protect health services (WHO 2022).

In the remainder of this section, we will present some security incidents in e-health systems. These security incidents will be presented based on the type of attack carried out.

2.5.1. *Example of a phishing attack*

The CHU[2] in Montpellier was the victim of a phishing attack in 2019. A CHU employee received an e-mail containing a virus. The virus infected over 600 computers at once. Fortunately, the WiFi network was not infected, enabling nursing staff to continue their medical procedures.

2.5.2. *Examples of ransomware attacks*

In May 2017, the WannaCry attack hit the British public healthcare system. By exploiting a Windows flaw, hackers managed to affect at least 16 health centers and 200,000 computers, leading to the cancellation of almost 20,000 consultations and the paralyzation of more than 1,200 pieces of diagnostic equipment.

In September 2020, Germany announced the first known death directly linked to a cyber-attack on a hospital. The hacking of the Düsseldorf hospital prevented the emergency care of a 78-year-old patient, who died after having to be sent to a more distant town. Investigators speculated that a group of Russian hackers might have been involved, trying to force the hospital to pay a ransom.

2 CHU: *Centre Hospitalier Universitaire*; France's university teaching hospitals.

In May 2021, the Irish public health service, HSE Ireland, was forced to shut down its entire IT system due to a "significant" cyber-attack using ransomware. The Irish Department of Health was subsequently also subject to a similar cyber-attack attempt, forcing it to suspend its IT system and leading to hospital appointment cancellations.

2.5.3. *Examples of data theft attacks*

In the summer of 2020, a public health establishment was hacked, and the personal data of 1.4 million French people who had taken Covid tests were stolen by the file storage and sharing service managed directly by this establishment. The attack was confirmed by Assistance Publique–Hôpitaux de Paris (AP-HP)[3]. The stolen data concerned patients' personal information, as well as the results of their Covid tests. The stolen information was accessed via a download platform in New Zealand.

In Finland, an attack on mental health services caused consternation in October 2020. Thousands of patient files, including those of children, are stolen from a private company that runs 25 psychotherapy centers (centers that manage people with psychological problems).

On April 23, 2021, the French newspaper *Libération* alerted pharmaceutical laboratories in France to an attack on their data. A file containing the confidential medical data of almost 500,000 patients is said to be in circulation with hackers. The file was created by hacking into the computer data of some 30 laboratories.

Another historic attack hit the hospital in Dax, in the Landes region, in 2021. Attackers were able to steal 10 years' worth of medical and personal data on patients.

In August 2022, the Corbeil-Essonnes hospital (serving 700,000 inhabitants in the Paris region) was hit by a cyber-attack, with a ransom demand of $10 million, later reduced to $1 or $2 million, depending on sources. The hackers had set an ultimatum of September 23 to pay the ransom. Once the deadline had passed, they released a series of data onto the black market.

3 It is the university hospital trust operating in Paris and its surroundings. It is the largest hospital system in Europe and one of the largest in the world.

2.5.4. *Examples of DDoS attacks*

In March 2020, Brno University Hospital in the Czech Republic fell victim to a large-scale DDoS attack in the midst of the Covid-19 pandemic. This attack caused the closure of three hospital departments for several weeks. As a result, patients were transferred to other hospitals and several surgical procedures were cancelled.

Another DDoS attack took place in March 2020, targeting part of the AP-HP's IT servers. The servers were overwhelmed by useless requests, blocking access to e-mails and remote working tools for telecommuting employees.

On March 16, 2020, the U.S. Department of Health website was also rendered inaccessible by a DDoS attack.

2.5.5. *Example of an internal attack*

In April 2018, U.S.-based health insurer Independence Blue Cross mistakenly uploaded a file containing the personal and medical data of nearly 17,000 patients. After launching an investigation in July, the company determined that an employee had mistakenly uploaded a file containing insurance member information to a publicly accessible website. The document was accessible for 3 months.

2.6. Existing security measures for e-health systems

In order to overcome the various problems of attacks and intrusions in e-health systems, a range of security mechanisms and measures need to be put in place. In the literature, several research studies have addressed this issue and proposed security measures for cloud- and IoT-based e-health systems (Butpheng et al. 2020). These security measures for e-health systems can be classified into five layers (virtually), where each of these layers must provide certain services and use the services of the previous layer to function properly (Butpheng et al. 2020). Let us take a closer look at each of the five layers proposed by Butpheng et al.:

– Object layer: this layer is where all the devices perform their specific functions, such as detection, monitoring, control and action, and is the bridge that connects the collected information to the control network. This bridge transmits information from the lowest to the highest layer of the IoT architecture. Connections to physical devices, such as wireless sensors, monitors and controls, can be direct or indirect. This layer is thus designed to prevent attacks, including physical attacks, identity theft, whitelisting, sandboxing, secure boot and sniffing.

– Communication/service layer: this layer enables different devices to interconnect on a central network to collect and share data on local devices. The transmission of signals between the layers uses various protocols, such as mobile/wireless, networks, Bluetooth and wired serial protocols. This layer is designed to provide secure communication, protected against sniffing and data tampering.

– Network layer: this layer is where IoT services and devices can connect to or be operated by cloud servers running off-site. This connection is established via physical hardware, such as switches, gateways, routers and broadband Internet. The network layer ensures that reliable response mechanisms, storage and analysis resources can be accessed and used by users, objects or services. This layer is designed to secure the routing and switching of data traffic against attacks such as man-in-the-middle, sniffing, spoofing and denial of service.

– Cloud layer: this layer provides distributed, Internet-based computing that can process, store and monitor the data collected by the object layer. Cloud computing services enable the use of IoT devices in the healthcare system, due to the ability to process massive amounts of data, almost instantaneously, in a scalable and low-cost way. In addition, cloud computing offers users data resources for computation, bandwidth, storage and services on a pay-per-use basis. This layer is designed to protect stored and processed data against unauthorized user access, tampering and attacks such as injection attacks, sniffing, cross-site scripting (XSS), phishing and viruses/malware.

– Application server layer: this layer is where users interact directly with the cloud-based IoT. Users can receive alerts, view collected data in real time, react manually/automatically to emerging situations, etc. This layer provides the user with application-specific services. It defines the different applications in which the IoT can be deployed, such as smart homes, smart cities and e-health systems. This layer is designed to protect against attacks such as information corruption and malicious code.

It should be noted that two transverse layers are also required according to this architecture (Butpheng et al. 2020):

– Transversal privacy layer: in this layer, respect for patient privacy is actively taken into account throughout the data lifecycle. Appropriate internal policies enable access to, or consultation of, personal data.

– Transversal security layer: this layer is where the system implements its security by considering security requirements from the physical layer to the application layer, such as security protocols, security transmission, secure inter-layer communication and data encryption, to prevent unauthorized access to systems and protect all data throughout its lifecycle by encrypting it. In addition, communication encryption, accountability tracking and data anonymization will be implemented at this level.

After presenting the various layers of e-health system security, we now turn to an annual survey conducted by the Healthcare Information and Management Systems Society (HIMSS 2021). In this survey, respondents are asked to specify the extent to which they have implemented a set of security controls within their healthcare establishments. These security controls can be diverse and include antivirus, firewall, encryption, multi-factor authentication, IDS (intrusion detection system), or IPS (intrusion prevention system). In what follows, we will take a closer look at some of the mechanisms shown in Figure 1.4.

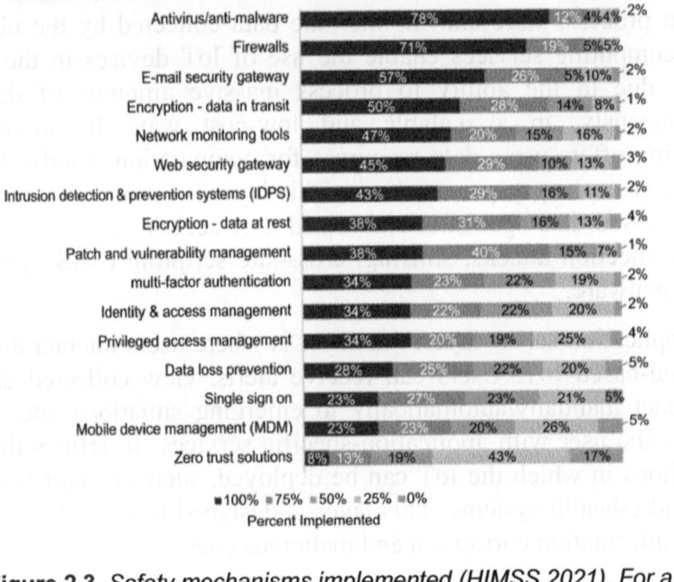

Figure 2.3. *Safety mechanisms implemented (HIMSS 2021). For a color version of this figure, see www.iste.co.uk/hamdi/ehealth.zip*

Let us start with the cryptographic solutions available to manage the sharing of data and storage of patient information across a network, cloud environment or via remote patient monitoring systems. We are going to focus on two solutions: homomorphic encryption and blockchain.

Homomorphic encryption (HE) guarantees enhanced security and confidentiality, while enabling the analysis of encrypted data and sensitive media information. Homomorphic encryption is versatile but has substantial computational requirements that currently slow down processing considerably. It can also be used in mobile devices to transfer and store medical data without decrypting it, preserving confidentiality if a node is compromised.

The second solution is blockchain, a peer-to-peer distributed ledger technology initially used in the financial sector. Its characteristics of decentralization, verifiability and immutability enable it to securely store personal medical data. Blockchain can solve the problem of patient data being used or shared without authorization via smart contracts for patient-regulated access to patient data. In this way, a patient can share their medical data on their own terms. Records can be kept anonymous or completely transparent for research purposes, depending on the patient's needs (Sarkar et al. 2021).

A second mechanism that can be implemented in e-health systems is multifactor authentication platforms. These help healthcare institutions secure remote access to confidential health information in a cost-effective and scalable way, without disrupting providers' workflow (Gonzalez-Granadillo et al. 2021).

2.7. Recommendations for protecting e-health systems

A number of recommendations can be used to alleviate any security problems that may affect e-health systems. These can be classified according to the type of action they propose: risk management methods, technical and organizational recommendations, and awareness and training initiatives.

2.7.1. *Risk management methods*

Tomashchuk (2020) has proposed a new threat and risk management (TRM) framework. It is based on the STRIDE and LINDDUN methodologies and uses components on feature space modeling, risk-based scoring, configuration decision support and regulatory compliance. The current proposal for this framework involves the following four steps:

– Model the variability of system characteristics: this involves covering a space of all possible features that the system architecture may possess and generating all valid feature configurations of these.

– Carry out threat assessments from the point of view of confidentiality and privacy: during this stage, risk extraction of identified threats also takes place. Given the input from the first step, risk assessment and extraction take place for each valid configuration.

– Generate privacy and confidentiality scores for each configuration and feature, based on specific metrics and the results of the previous step.

– Find the most appropriate solution (both in terms of privacy and confidentiality) from the point of view of the architecture configuration that meets a certain risk budget and satisfies minimum requirements.

2.7.2. *Technical and organizational recommendations*

As shown in Table 2.1, the Swiss National Center for Cybersecurity (NCSC 2022) has defined minimum technical and organizational requirements for cybersecurity in the healthcare sector. For each of the measures listed in this table, the level of implementation (organizational and/or technical) is indicated, as well as the nature of the guideline (mandatory or optional).

Every organization is required to develop and maintain a patch and software lifecycle management plan. A software and security update (patch) management system can be used to manage patches and the software lifecycle. Software and devices included in the security perimeter (antivirus, firewalls, proxy servers or IDS/IPS intrusion prevention systems, etc.) must record all activities. These activities must be immediately monitored to identify any suspicious situation, intrusion attempt or attack.

An organization's internal resources accessible via the Internet must be secured by a second factor. Back-up copies of data must be kept secure and separate from the network to ensure that when an organization is the victim of a ransomware attack and its data have been encrypted, there is a working backup copy that can be restored once the infected elements have been netted.

Measures	Level of application	Nature of the directive
Technical and organizational patch and lifecycle management	Organizational	Obligatory
Close monitoring of data concerning the security perimeter	Organizational and technical	Obligatory
Lose monitoring of terminals	Technical	Optional
Patch and lifecycle management	Organizational and technical	Obligatory
Integration into NCSC's closed circle of customers	Organizational and technical	Optional
Offline backup/disaster recovery	Technical	Obligatory
Network segmentation	Technical	Obligatory
Authentication protection	Technical	Obligatory
Blocking of dangerous e-mail attachments	Technical	Optional
File execution control	Technical	Optional

Table 2.1. *Overview of the measures described (NCSC 2022)*

2.7.3. *Raising awareness and training*

Security awareness training for the healthcare sector is more important than ever. Indeed, healthcare organizations are increasingly finding themselves on the front line of cyber-attacks involving malware, security incidents, phishing, DDoS and data theft. It is clear that hackers will continue to target the healthcare sector as long as there is profit to be made, whether it is selling stolen patient data or holding e-health systems hostage until their demands are met.

Awareness-raising and/or training initiatives can empower employees to change their behavior, identify threats and protect your organization from risk.

A recent study (Hyla and Fabisiak 2020), carried out in the context of the digitization process in the Polish healthcare sector, led to the following conclusions:

– A lack of standardized applications and the distributed nature of medical records, with only a few central government applications, make it difficult to design a short, comprehensive cybersecurity training course for all workers.

– Many healthcare sites use software with obsolete security measures, or without adequate security measures, so users need to be better trained.

– Many healthcare professionals (especially older ones) have a problem using a computer.

As a result, cybersecurity training needs to be improved and applied more frequently to better:

– integrate simple practical examples;

– incorporate real-life demonstrations of past cyber-attacks that have been professionally managed;

– incorporate role-playing activities that explain the persuasive techniques used by cybercriminals;

– make training updates periodic (e.g. once a year).

Cybersecurity awareness campaigns, in general, need to take into account the factors that influence human behavior (Dolan et al. 2010). Among other things, training must be aligned with risks and designed to change people's behavior by providing simple, consistent and easy-to-follow rules of behavior (Bada et al. 2015). Bada et al. (2015) offer a good analysis of these factors and analyze some generalities about cybersecurity awareness campaigns. In addition, the use of various teaching methods enables people to better understand a given topic (Abawayj 2014).

2.8. Conclusion

e-Health systems are becoming increasingly widespread around the world, creating huge potential to improve clinical outcomes and transform care delivery. However, there are growing concerns about the security of healthcare data and devices. Increased connectivity has exposed medical devices to new cybersecurity vulnerabilities. Healthcare has become an attractive target for cybercrime for two fundamental reasons: it is a rich source of valuable data and its defenses are weak.

In addition, the medical ecosystem is complex and undergoing rapid digital transformation. Subject to numerous regulations, it is vulnerable and requires vigilance. Everything remains to be done, and the challenge is to provide a response in line with the transition and modernization that the sector is undergoing. Indeed, cybersecurity breaches are very diverse and include theft of healthcare information, ransomware, attacks on hospitals and potential attacks on implanted medical devices. Breaches can reduce patient confidence, cripple e-health systems and threaten human life.

Security measures must also be put in place, to minimize risks, such as detecting/recognizing vulnerabilities, and, more generally, mapping risks, drawing up a cybersecurity strategy including the definition of roles and responsibilities, response in the event of an attack and so on. It is also important to take advantage of blockchain in e-health, and biometric systems, both to secure communications and for authentication, not forgetting the implication and impact that artificial intelligence can have for e-health security.

2.9. References

Abawajy, J. (2014). User preference of cyber security awareness delivery methods. *Behaviour & Information Technology*, 33(3), 237–248.

Agence France-Presse (2022). Cyberattaques contre hôpitaux et systèmes de santé : un phénomène mondial [Online]. Available at: https://www.lesoleil.com/2022/11/09/cyberattaques-contre-hopitaux-et-systemes-de-sante-un-phenomene-mondial-dc29d3a71d5a944bfb02d57ecfbc7737.

Bada, M., Sasse, A.M., Nurse, J.R.C. (2015). Cyber security awareness campaigns: Why do they fail to change behaviour? *International Conference on Cyber Security for Sustainable Society*.

Bhambere, S., Abhishek, B., Sumit, H. (2021). Rapid digitization of healthcare – A review of COVID-19 impact on our health systems. *International Journal of All Research Education and Scientific Methods (IJARESM)*, 9(2), 2455–6211. doi: 10.6084/m9.figshare.14156417.v1.

Butpheng, C., Yeh, K.H., Xiong, H. (2020). Security and privacy in IoT-cloud-based e-health systems – A comprehensive review. *Symmetry*, 12(7), 1191.

Centre national pour la cybersécurité (2022). Cybersécurité dans le secteur de la santé : recommandations [Online]. Available at: https://www.ncsc.admin.ch.

Chng, S., Lu, H.Y., Kumar, A., Yau, D. (2022). Hacker types, motivations and strategies: A comprehensive framework. *Computers in Human Behavior Reports*, 5, 100167.

Dolan, P., Hallsworth, M., Halpern, D., King, D., Vlaev, I. (2010). MINDSPACE influencing behaviour through public policy. Report, Institute for Government, Cabinet Office.

Gonzalez-Granadillo, G., Diaz, R., Veroni, E., Xenakis, C. (2021). A multi-factor assessment mechanism to define priorities on vulnerabilities affecting healthcare organizations. *ITASEC'21: Italian Conference on Cybersecurity*, Rome.

HIMSS (2021). 2021 HIMSS healthcare cybersecurity survey [Online]. Available at: 2021_himss_cybersecurity_survey.pdf.

Hyla, T. and Fabisiak, L. (2020). Measuring cyber security awareness within groups of medical professionals in Poland. *Proceedings of the 53rd Hawaii International Conference on System Sciences*.

Kavallieros, D., Myttas, D., Kermitsis, E., Lissaris, E., Giataganas, G., Darra, E. (2021). Using the dark web. In *Dark Web Investigation. Security Informatics and Law Enforcement*, Akhgar, B., Gercke, M., Vrochidis, S., Gibson, H. (eds). Springer, Berlin. doi: 10.1007/978-3-030-55343-2_2.

Neumann, M., Fehring, L., Kinschez, K., Truebel, H., Dahlhausen, F., Ehlers, J.P., Mondritzki, T., Boehme, P. (2021). Perspective of German medical faculties on digitization in the healthcare sector and its influence on the curriculum. *GMS J Med Educ*, 38(7). doi: 10.3205/zma001520.

Organisme mondial de la Santé (2022). Système de surveillance des attaques contre les soins de santé [Online]. Available at: https://extranet.who.int/ssa/Index.aspx.

Patru, L.G. (2022). Stakeholders perception of the importance of digitalization in healthcare organizations. *Management and Marketing Journal*, 2, 277–285.

Sarkar, A., Maitra, T., Neogy, S. (2021). Blockchain in healthcare system: Security issues, attacks and challenges. In *Blockchain Technology: Applications and Challenges*, Panda, S.K., Jena, A.K., Swain, S.K., Satapathy, S.C. (eds). Springer, Berlin, 113–133.

Seh, A., Zarour, M., Alenezi, M., Sarkar, A.K., Agrawal, A., Kumar, R., Khan, R.A. (2020). Healthcare data breaches: Insights and implications. *Healthcare (Basel)*, 8(2), 133. doi: 10.3390/healthcare8020133.

Sütterlin, S., Knox, B.J., Maennel, K., Canham, M., Lugo, R.G. (2022). On the relationship between health sectors' digitalization and sustainable health goals: A cyber security perspective. *Good Health and Well-Being*, 133.

Tomashchuk, O. (2020). Threat and risk management framework for e-health IoT applications. In *Proceedings of the 24th ACM International Systems and Software Product Line Conference.*

Zeadally, S., Isaac, J.T., Baig, Z. (2016). Security attacks and solutions in electronic health (e-health) systems. *J Med Syst*, 40, 263.

3

Security Policies for e-Health Systems

Ryma ABASSI
Innov'COM, SUP'COM, Université de Carthage,
Tunis, Tunisia

3.1. Introduction

Security issues are of vital importance to the life and development of a system, and the concept of security policy (SP) has emerged as a key element in the process of securing systems. It is, in fact, through an SP that we can separate the actions that must be authorized from those that must not.

Today's organizations are increasingly aware of the importance of securing their IT systems and/or communication networks in order to guarantee fundamental security requirements, that is, confidentiality, integrity and availability. Nevertheless, any security solution implemented in a given organization without prior analysis of its real needs can cause damage. To avoid such problems, it is a good idea to define a set of rules separating authorized and unauthorized actions before implementing any security solution. Once defined, these rules are grouped together in a document called a "security policy", which is intended to form the basis of any security solution.

Given the importance of an SP in securing an organization, it is imperative that an SP be provided with an adequate framework for specifying, manipulating, validating and implementing it with the utmost

e-Health Security Management,

care, and for a maximum reduction in risks and malfunctions. More specifically, we believe that to achieve such a goal, it is necessary to be able to associate adequate representation, validation, verification and testing techniques with an SP.

This is all the more true for a critical domain such as healthcare. Indeed, the ever-increasing use of digital systems and applications in e-health systems, particularly those powered by modern information and communications technologies, has created a threat area in which policy and regulation are struggling to keep pace with development. Recall that one of the most famous healthcare cybersecurity breaches was the WannaCry malware attack worldwide, which crippled 600 UK healthcare organizations in 2017 (Ghafur et al. 2019). Recent successful cyber-attacks – for example, the ransomware attack on German healthcare group Fresenius in May 2020, or the Doctolib data theft in July 2020 – show that the risks are there.

To improve cybersecurity, healthcare providers and professionals need to implement the following practices: secure communication to prevent unauthorized access and modification, data protection and a strategy to control and limit access. This chapter therefore focuses on security policies in e-health systems.

The rest of this chapter is organized as follows: in section 3.2, the concept of SP is defined, along with its various inherent concepts. In section 3.3, we introduce the concept of executable SP, a concept used to facilitate the validation and testing of SPs. Section 3.4 presents security services in e-health systems. Security requirements for e-health systems are dealt with in section 3.5. Some future security challenges for e-health systems are presented in section 3.6. Finally, section 3.7 concludes this chapter.

3.2. The concept of the security policy

In this section, we define the concept of SP and introduce its main concepts.

3.2.1. *Definition*

Defining the concept of the SP is a rather delicate task, given that it can be used to refer to various aspects of safety. Various definitions have been proposed in the literature. According to RFC 2196 (Fraser 1997), "A security policy is a formal statement of the rules by which people who are given access to an organization's technology and information assets must abide". This definition covers all the SPs that govern the behavior of applications running in an environment that we wish to secure. An example of such an SP would be: "An application is forbidden to send information from a file in the 'Secret' directory".

Sterne (1991) proposes a "three-step" definition for SPs, in which he highlights three fundamental and complementary elements: the SP objective, the organizational SP and the automated SP, which can be detailed as follows:

– The objective of the SP is to declare the intention to protect a resource against unauthorized use. This declaration must identify which uses are permitted and which are not. Once the objective of the SP has been stated, the next step is to provide the necessary means to achieve it. These means are spelled out in the organizational SP.

– The organizational SP is a set of laws, rules and practices that govern how to manage, protect and, where appropriate, allocate resources to achieve SP objectives. To be effective, these laws, rules and practices must give individuals the ability to know whether, or not, their actions violate the SP. Furthermore, a refinement of organizational SP is needed to take into account the security of information manipulated through communication networks in order to establish what is considered "acceptable behavior". An automated SP corresponds to this refinement.

– The automated SP is a set of restrictions and properties that specify how a communication network prevents the misuse of its resources to violate the organizational SP.

For the purposes of this chapter, we adopt the definition given in RFC 2196 (Fraser 1997). The advantage of such a definition is that it is generic, in the sense that it is valid for most communication contexts. On the other hand, the definition of an SP must be related to the context in which it is implemented. In what follows, three examples relating to three types of SP are presented (Abassi and Guemara El Fatmi 2019):

– Network SPs: network infrastructures are often used to carry out a variety of attacks (such as flooding and e-mail spoofing). This is why it is necessary to prevent network nodes from passing unauthorized traffic. Network SP therefore consists of a set of rules specifying how data are to be transmitted across the network. It comprises two components: preventive network SP and reactive network SP. The former attempts to mitigate vulnerabilities so that attacks do not occur, while the latter attempts to limit the damage resulting from the occurrence of a security incident.

– Access control SP: because of the multitude of security threats exploiting operating system vulnerabilities, a set of security mechanisms must be implemented at this level. This set of mechanisms, known as the trusted computing base (TCB), concerns the resources of the communications network (e.g. hardware, software and processes).

– Key management SPs: in order to establish a secure tunnel using the IPSEC protocol, two endpoints must agree on a set of cryptographic parameters. These parameters are grouped together under the term "Security Association" (SA) and are managed according to the local SP at each end node.

More generally, an SP serves to implement the principles of an organization's security strategy, and to inform users, staff and managers of their imperative obligation to protect technological and informational assets (Fraser 1997). Therefore, an SP must first be able to be implemented through system administration procedures by publishing guidelines for acceptable use. It must then be implemented with the appropriate security tools, and finally, it must clearly define the areas of responsibility of users, administrators and managers.

In addition, a "good" SP should be based primarily on the following nine components (Fraser 1997):

– IT procurement guidelines that specify the required (or preferred) security features. These should be in addition to existing SPs and acquisition guidelines;

– a confidentiality policy that defines reasonable expectations of confidentiality regarding issues such as e-mail monitoring, logging and access to user files;

– an access policy that defines the access rights and privileges used to protect assets against loss or disclosure, through the specification of acceptable usage guidelines for users, operations personnel and managers;

– an accounting policy that defines the responsibilities of users, operating staff and managers. This policy should specify audit capabilities and provide guidelines for countermeasures;

– an authentication policy that establishes trust through an effective password policy, guidelines for authenticating remote locations and the use of authentication equipment (such as one-time passwords and the equipment that generates them);

– an availability statement that manages users' expectations regarding resource availability. This declaration must address issues of redundancy and recovery, as well as the specification of operating hours and downtime for maintenance;

– an IT system and network maintenance policy that describes how maintainers, both internal and external, are authorized to handle and access technology. An important issue here is whether remote maintenance is permitted, and how such access is managed;

– a breach reporting policy that specifies the type of breaches to be reported and to whom;

– support information that provides users, staff and managers with contact details for each type of policy violation. This information provides guidelines on how to handle external investigations into a security incident.

The environment we wish to begin building, and which will be proposed in this book, attempts to take into account the definition we have just cited, as well as its main components. The cornerstone of this environment is based on the modeling of an SP.

3.2.2. *Modeling a security policy*

Broadly speaking, a model can be defined as a formalism that offers a subjective but relevant view of reality. A model is used to better understand the system being developed, that is to visualize its properties, specify its structure or behavior, document and guide its construction, evaluate its performance, etc.

Therefore, an SP model can be defined as a formalism that enables the SP to be represented in a clear and unambiguous way. It is a formal expression of an SP, which can be used to describe and possibly verify the safety of a network. In addition, an SP model can be used to formalize security requirements relating to the confidentiality, integrity, or service availability properties of a communication network.

Most of the work on SP modeling has focused on access control models. These specify the mechanisms to be implemented within the network in order to guarantee the expected security services. The model structure is often based on an abstract finite-state machine. In what follows, we will review some of these models, in particular discretionary access control (DAC) models, mandatory access control models and role-based access control models. Role-based access control models have recently been introduced for commercial applications and are seen as an alternative to discretionary and mandatory models.

3.2.2.1. *Discretionary access control model*

A DAC model is the formalization of a discretionary SP. An SP is said to be discretionary when the management of access to data belonging to a subject is at the latter's discretion: each subject can specify, for objects belonging to it, the access rights granted to other subjects. On the other hand, according to a discretionary SP, a subject possessing a certain access right is able to confer this right on any other user (Kashmar et al. 2021). The management of access rights is generally given by the following rule: "A subject *s* can give an access right on an object o to another subject *s* if and only if *s* is the pro-owner of the object *o*".

However, the main problem with DAC is that it requires trust in the system's users, since information security is not a feature of the communication network, but a function of the users. Furthermore, this approach does not guarantee that access to an object is forbidden to a subject: it is sufficient for the subject to obtain, in one way or another, the right to access the object in order to be authorized to access it. DAC SPs can do nothing to prevent the use of hidden storage channels, as used by Trojans in particular. Finally, a DAC SP focuses exclusively on the execution of object operations, that is, accesses to objects, independently of the flow of information between objects resulting from each access. It is this limitation that mandatory access control SPs attempt to overcome.

3.2.2.2. *Mandatory access control model*

A mandatory access control (MAC) model is a formalization of a mandatory SP. An SP is said to be mandatory when it manages access to objects based on the level of sensitivity of the information contained in the objects. Therefore, the rule of a mandatory access control SP can be expressed as follows (Abassi and Guemara El Fatmi 2019): "A subject is authorized to access an object if and only if the subject is authorized to access the information contained in the object". A mandatory SP is generally used as a complement to a DAC SP; the rules of the mandatory access control SP are considered mandatory constraints, and each user can specify their own access rights for the objects belonging to him. For example, the Bell and LaPadula model (Cristiá and Rossi 2021), also known as the BLP model, is a mandatory access control model that extends the discretionary model. It is one of the most effective models for implementing confidentiality in a system. In this work, we adopt a minimalist approach and essentially present the basic properties of this model. The BLP model is built on the concepts of object, subject, access rights and access matrix, expressing the state of a network in terms of access rights. In this model, the emphasis is on controlling the flow of information between objects and subjects, rather than just controlling access to them. Therefore, the BLP model allows four modes of access: read-only, write-only, execute and read/write. Formally, the system is modeled by an abstract finite-state machine defined by a triplet (V, R, T), as presented above.

Given S the set of subjects, O the set of objects, A the set of access properties = {*execute, read, append, write*} and L the set of security levels with partial order ≤, a state is formally defined by:

$$(b, M, f) \in B \times M \times F$$

where:

– B = P(S×O×A) is the set of current access operations;

– M is the set of access permission matrices M = (Ms, 0) s∈S, o∈O;

– F⊂ LS × LS × LO is the set of safety level assignments. An element *f* of F is triplet (*f*s, *f*c, *f*o), such that:

– *f*s: S→ L gives the maximum level a subject can reach;

– *f*c: S→ L gives the current level of each subject;

– *f*o: O→ L gives the classification of each object.

3.2.2.3. *Role-based access control model*

Access control models such as DAC and MAC can only model SPs that are restricted to static permissions. They do not offer the possibility of expressing contextual rules relating to permissions, prohibitions, obligations and recommendations. The role-based access control (RBAC) model was introduced (Kashmar et al. 2021). In this model, SP is not applied directly to users, as in previous models, but to roles. Therefore, permissions are granted to roles and users obtain the permissions granted to the roles they play. However, RBAC suffers from certain limitations, including the impossibility of expressing permissions, and in particular context-dependent permissions. As a result, it is difficult to specify that a doctor only has permission to access a patient's medical record if the patient is their patient. In addition, the concept of role hierarchy is somewhat ambiguous, and the distinction between role and group is blurred.

To address these limitations, the organization-based access control (Or-BAC) model has been proposed (Kashmar et al. 2021). Or-BAC takes the basic concepts of RBAC and supplements them with another key concept, that is, of the organization. An organization is any entity that is responsible for managing a set of security rules (permissions or prohibitions). For example, a particular hospital is an organization. A security component, such as a firewall, can also be assimilated to an organization, insofar as it manages a set of security rules. Therefore, the definition of a role in the Or-BAC model is always associated with a particular organization. Furthermore, the Or-BAC model takes into account rules specifying permissions or contextual prohibitions, as well as those specifying obligations and recommendations.

3.2.2.4. *Key features required of an SP model*

Existing SP models seem insufficient to solve the specificities of SP validation, verification, or testing. Discretionary models have serious drawbacks, particularly with regard to information leakage for unauthorized subjects, since they do not control the information flows acquired by a subject. Compulsory SP models, on the other hand, are highly rigid and very specific to the domains for which they were developed. What is more, they are too centralized and ill-suited to networked analysis and decision-making, and do not easily take into account confidentiality, integrity and, a fortiori, availability requirements. The Or-BAC model seems to be the most comprehensive to date, but it is rather complex, and some concepts are somewhat ambiguous. Such is the case with the concept of role hierarchy, as it is not clear that the latter corresponds to the network hierarchy.

We have also identified various criteria that need to be taken into account when designing an SP model, including:

– Ease of use and comprehension: an SP model must be sufficiently simple and comprehensible to facilitate SP implementation.

– Completeness, to support the specification of different types of rules (authorizations, obligations, prohibitions and delegations): an SP model must be sufficiently complete, taking into account all the concepts required to implement any type of SP in any environment.

– Expressiveness, to enable the analysis of SPs for inconsistencies in the specification, for example: an SP model must not only facilitate the specification of the SP but must also form the basis for any analysis of the latter, by providing for processes such as validation, verification or testing.

– Adaptation to the requirements of the safety domain, such as constraints and events: an SP model must not only take into account the basic concepts required to implement an SP but must also be capable of implementing all the specific requirements that may be used in such a domain.

3.3. Environment for specifying, validating and testing security policies

In recent work (Abassi and Guemara El Fatmi 2019), we have proposed an environment for specifying, validating and testing SPs based on the concept of executable security policy (ESP). Therefore, we introduce the ESP concept as a model that allows to generate the behavior of the system interacting with its environment according to an SP. This can be achieved through the use of an executable language for specification. An SP specification, although formally defined, may contain errors due to human intervention in the specification, which is prone to omissions, errors, misunderstandings, etc. To avoid this kind of mishap, we have developed a "model" that can be used to generate the behavior of the system interacting with its environment. To avoid such problems, it is essential to validate the SP before implementation. Validating an SP means checking that it meets the desired security requirements. The executable nature of a PSE can be used to validate an SP.

3.3.1. *Specifying a security policy*

Using the definitions set out above and knowing the nature of the global network in which the SP must work, we propose the following modeling based on the following concepts:

– Subjects: these represent active entities in the system, such as users, employees, processes and applications.

– Objects: these represent passive entities in the system (also known as assets or resources), such as ports and documents.

– Actions: these represent the actions that a subject can perform on an object, such as writing and reading.

– Security rules: these express the security decisions appropriate to each situation. Two decisions are envisaged in an SP: either a permission or a refusal.

– Constraints: these represent conditions that must be met before an action can be carried out, and generally limit the scope of security rules.

– Events: their occurrence triggers a rule involving a given subject, object and action.

Let $s \in S$ be the set of subjects, $o \in O$ the set of objects, $a \in A$ the set of actions, $c \in C$ the set of constraints, $e \in E$ the set of events and $resp \in \{yes, no\}$ the SP response. Note that constraints and events are optional arguments that are listed when they exist and whose absence is represented by "–".

These concepts are used to model four types of rules as follows:

– Authorization rules distinguish between authorized and prohibited actions. This type of rule can be seen as a request for which a response must be given. Formally, we express them as:

$$req(s \times o \times a \times [c] \times [e]) \rightarrow resp$$

Note that the resp value can change over time depending on whether or not the rule's constraints are satisfied, and that a positive response implies a positive authorization, while a negative response implies a negative authorization.

– Obligation rules express actions that subjects must perform in response to a triggering event. Formally, we express them as:

$$ob\ (s \times o \times a \times [c] \times e)$$

For example, ob (user, change-pwd, account, -, start of month) means that every “user” must “change the password” of their “account” at the “start of month”.

– Prohibition rules express actions that subjects must never perform. Such rules are similar to negative authorization rules, with an always negative response. Formally, we express them as:

$$\text{phb}\ (s \times o \times a \times [c] \times [e]) \rightarrow \text{no}$$

For example, a patient trying to access another patient’s file will always be denied access.

3.3.2. *The concept of executable security policy*

In line with the executable specifications used in software engineering, and in order to see the actual behavior of a specification, we define a PSE as a model of the SP capable of generating the behavior of the system to be secured interacting with its environment, according to the security requirements specified by the SP. In this way, by using a PSE, the behavior of the SP can be observed and tested before it is implemented in the network, thus avoiding a great deal of inconvenience.

Figure 3.1 depicts an SP as a means of communication between two network components, for which communication takes place according to the various rules forming the SP. In this figure, four entities are described: the subject, the object, the SP and the event trigger, which contains a set of events external to the communication network that are triggered at particular times, known and/or programmed in advance, and which lead to the execution of one or more actions. The SP is an essential part of any interaction between subject and object. It should be noted that SP can have two types of activity. The first is verification, in the sense that the SP verifies any traffic attempting to pass through it in the subject-to-object direction and can then either allow or forbid this passage. The second activity, on the other hand, is a transmission without verification, which concerns the implementation of traffic passage in the object-to-subject direction in response to an action request already sent.

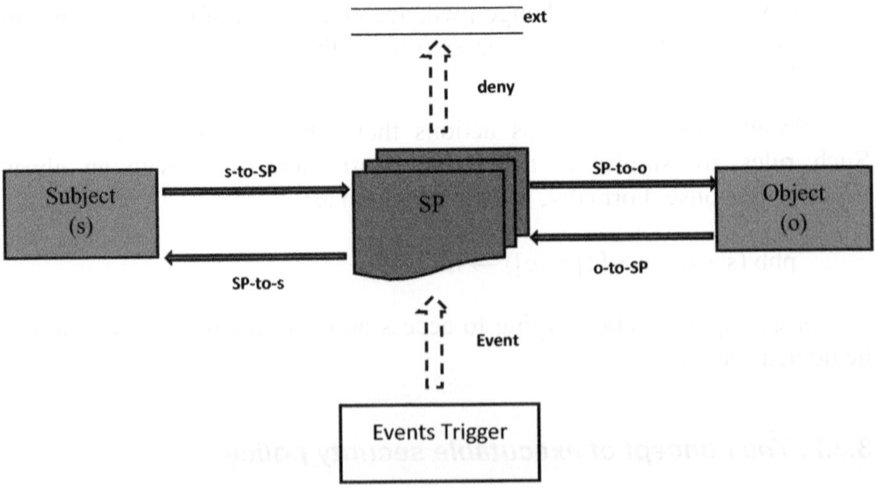

Figure 3.1. *Modeling an ESP*

3.3.3. *Testing a security policy*

We propose a formal environment for test case generation based on system modeling, SP specification (test goals) and the use of a model checker. Once generated, these test cases need to be improved in order to quantify their effectiveness in detecting flaws. This is done by generating mutants. In fact, the use of policy mutants as surrogates for real faults enables an initial investigation of SP testing. Four steps are required: (1) the SP is specified via a new executable modeling language; (2) test objectives are expressed using LTL temporal logic; (3) test cases are generated using a classical model checking technique; and (4) the generated test cases are improved by adapting mutation analysis and defining mutation operators.

Test objectives are designed to represent the functionalities of system behavior. Therefore, we associate a test objective with each rule type and define them by LTL according to what is required by the model checker. The test objective can be expressed with LTL using Boolean operators and F and G temporal modalities. A test objective is expressed by a logical formula based on literals.

A test case is defined as an elementary test aimed at testing a particular functionality (Abassi and Guemara El Fatmi 2019). This step is carried out by a classical model checking technique using a model checker: given the system model and the negation of TP, the model checker produces a counterexample showing a trace where the negation of the test objective does not hold and, consequently, where the desired objective does. This counterexample is the desired test case. In addition, two verdicts are associated with a CT: a fail verdict denotes a divergence from the expected behavior, and the ESP is rejected. A pass verdict is returned if the observation is correct, and the objective is achieved.

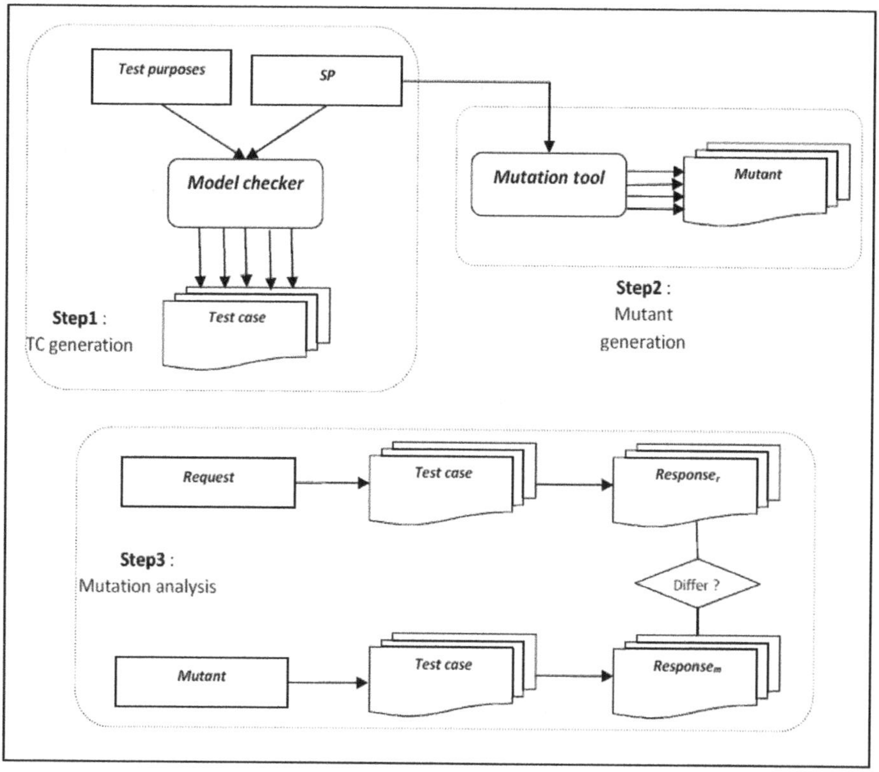

Figure 3.2. *Security policy test environment*

Figure 3.2 describes the proposed environment. In Step 1, test cases are generated, as described above. In Step 2, mutants are generated using the SP under test and a mutation tool implementing certain mutation operators. In Step 3, these mutants are applied to the test cases and the resulting response, response$_m$, is compared with the response of these test cases when prompted by the user's initial requests response$_r$. If these responses are the same, then the mutant is killed; otherwise, it is alive.

3.4. Security services for e-health systems

Data security has an impact on patient care. Inability to access data could lead to delays in treatment and decision-making. In 60 mHealth applications analyzed, 137 security vulnerabilities were found, with remote monitoring applications presenting the most vulnerabilities (32.12% of the total) (Beltran et al. 2015). The risk factors established by OWASP were taken into account (OWASP 2022); 64% of the vulnerabilities found corresponded to "security decisions via untrusted inputs", meaning that the attacker elevated access and privileges, affecting the confidentiality and integrity of clinical data.

3.4.1. *The e-health concept*

According to the World Health Organization (WHO), e-health is defined as "digital services for human well-being", that is, the application of information and communication technologies to health and well-being. Some even associate the "e" with several "e" (Eysenbach 2001):

– Efficiency: one of the promises of e-health is to increase the efficiency of healthcare, thereby reducing costs. One possible way of reducing costs would be to avoid redundant or unnecessary diagnostic or therapeutic interventions by improving communication possibilities between healthcare facilities and involving patients.

– Evidence-based: e-health interventions must be evidence based in the sense that their effectiveness and efficiency must not be supposed but proven by rigorous scientific evaluation. Much work remains to be done in this area.

– Encouragement of a new relationship between patient and healthcare professional toward a true partnership, where decisions are made on a shared basis.

– Education of doctors via online sources (continuing medical education) and consumers (health education, consumer-friendly preventive information).

– Enabling information exchange and communication in a standardized way between healthcare institutions.

– Extending the scope of healthcare beyond its conventional boundaries. This is understood in both a geographical and a conceptual sense. e-Health enables consumers to easily obtain healthcare services online from global providers. These services can range from simple advice to more complex interventions or products such as pharmaceuticals.

– Ethics: e-health involves new forms of patient–physician interaction and poses new challenges and threats to ethical issues, such as online professional practice, informed consent, confidentiality and equity questions.

– Equity: one of the promises of e-health is to make care more equitable, but, at the same time, there is a considerable threat that e-health could widen the gap between the "haves" and the "have-nots". People who do not have the money, skills and access to computers and networks cannot use computers effectively. As a result, these patient populations (who would benefit most from health information) are those least likely to benefit from advances in information technology, unless policy measures ensure equitable access for all. The digital divide currently exists between rural and urban populations, rich and poor, young and old, men and women, and between neglected/rare and common diseases.

3.4.2. *Comparison of national digital health infrastructure security policies*

We have tried to review the different requirements described by different countries (ENISA 2015) and synthesize the result below.

In Belgium, security requirements specific to the e-health platform are imposed by law and concern access control, end-to-end encryption, exchange protocols and security measures. Specific measures are taken to prevent the centralized storage of personal health data, and to ensure the secure exchange of electronic data between all actors.

In Austria, healthcare is considered a critical sector, which means that national legislation applies to critical systems. Security measures are mandatory by law and linked to civil protection guidelines. In terms of data confidentiality, ELGA (*Elektronische Gesundheitsakte* (or electronic health record)) is an opt-out system in which a patient must formally request that their record not be made available via ELGA. Access to ELGA and to a patient's record is strictly limited to the attending physician and the patient. The physician uses the patient's smart card and must have a justified need to access the data. Other doctors, government agencies, payers (they are part of the government) and employers do not have access to patient data. Since ELGA is based on a federated architecture, no patient medical data are stored centrally. While this eliminates the possibility of any kind of analysis, it offers an additional level of security that hybrid and centralized architectures cannot match.

In Estonia, security systems are based on five principles:

– secure authentication of all users (biometric or knowledge-based user authentication, device authentication). The Estonian ID card is an effective security device. Its main task was digital signature, but the same mathematical method and device can be used successfully for every type of secure authentication;

– maximum accountability and transparency, with all actions leaving an immutable, irremovable security trail;

– separation of personal data from medical data, known as personal data encryption;

– database encryption to minimize the risk of confidentiality for technical administrators;

– effective monitoring of all actions and corresponding countermeasures, both organizational and technical.

Other security requirements include compliance with national regulations, laws and security standards, implementation of security measures, external auditors, certification to national or international standards, reporting of security incidents to the relevant authorities and incident management, segregation of duties, risk management, awareness and training, access control, network security and business continuity.

In Germany, as described in the recently passed IT security law, here are the specific requirements that should apply to all e-health services:

– design and implementation of specific safety standards;

– identification of critical infrastructures;

– incident reporting;

– mandatory audits and reports;

– collaboration and information sharing with operators.

The flow of information between the professional domain and the public domain needs to be monitored in order to control how data are used by healthcare professionals and to ensure the security of sensitive information for the public. The monitoring mechanism can be implemented in the form of an information accountability framework. The three main aspects are legal, social and technological, and their interrelationships are illustrated in Figure 3.3.

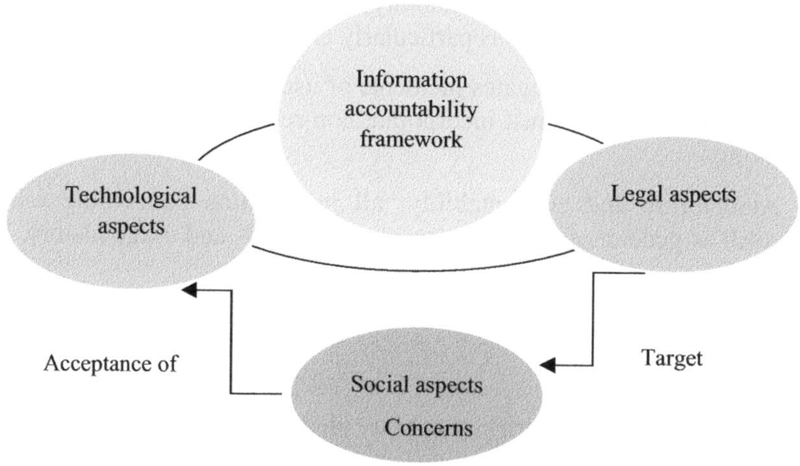

Figure 3.3. *Information accountability framework (Gajanayake et al. 2012)*

3.5. Security requirements for e-health platforms

Securing e-health data is a major challenge for two main reasons: the high computational overhead when encryption techniques have been used, and the sensitivity of personal medical information to change when modification techniques are used.

3.5.1. *Essential security functions*

e-Health platforms must ensure the following security functions (Khan et al. 2012; Oh et al. 2021; Ksibi et al. 2022):

– Privacy: in this kind of platform, this includes anonymity. Patients must not be identifiable by unauthorized users, such as management staff and insurers, but only by doctors and nurses.

– Confidentiality: ensures that only authorized entities can read data. In this context, authorizations must be granted with due respect for patient privacy.

– Integrity: checks that data received have not been altered.

– Authentication: ensures that equipment, applications and systems are mutually recognized when they wish to communicate.

– Availability: ensures the availability of services and data in the case of a denial-of-service attack. This is particularly crucial in emergency situations.

– Non-repudiation: mitigates incidents of user liability and/or negligence. Access control systems must incorporate non-repudiation mechanisms, such as auditing.

– Auditability and accountability: all information concerning e-health data, such as generation time, owner, access records and usage history, must be recorded.

3.5.2. *Security models*

Three levels of security can be identified: data collection, data transmission and data access storage (Zriqat et al. 2016).

3.5.2.1. *Security models at the data collection level*

In e-health systems, sensors are needed to collect medical data, whether in an environment where the sensors are isolated, such as those used to record blood pressure and ECG, or networked, such as BANs (body area networks). In both cases, the data collected are transferred to a processing unit, which may be a server or a mobile device. Various encryption techniques have also been proposed.

Garcia-Monchon et al. (2010) proposed a two-level extension of the RBAC model: (1) to assign and distribute access control policies to sensor nodes; (2) to store the current medical context (location, time, health information) that influences access control decisions on the patient's medical situation (critical, emergency or normal). The modular design makes system configuration more efficient and simplifies policy composition for deploying safer, more secure medical sensor networks. However, when an emergency or a critical situation arises, medical elements can override access restrictions to sensitive data that were restricted under normal conditions.

One of the limitations of this model is that there is no mechanism for detecting unauthorized access when critical situations arise.

Sawand et al. (2015) stated that there are often two types of sensors deployed in an e-health system: sensors placed on or in a patient's body and those deployed on hospital premises/integrated into certain equipment, such as a smart hospital bed. Essentially, such sensors can be compromised by an attacker, their cryptographic keys and status information can be acquired, and they can be cloned to install numerous malicious sensors and compromise the entire system. As solutions, symmetric key cryptography and public key cryptography have been widely used.

3.5.2.2. *Security models at the data transmission level*

Several attacks can target data transmission. Eavesdropping, denial-of-service and modification are the most significant attacks that can be launched in this context.

In Lin et al. (2009), a strong privacy-preserving Scheme Against Global Eavesdropping (SAGE) for e-health systems was presented. It ensures content-based confidentiality by applying cryptography. Based on the broadcast strategy, it ensures receiver anonymity and guarantees non-correlation, since an attacker cannot link a patient to a specific doctor. Indeed, it uses pseudonyms instead of the user's real identity. However, the anonymity of the sender is not guaranteed since direct communication between the patient and the online health center is not prevented.

The PEACE (patient-centric access control) scheme was introduced in Barua et al. (2011). To ensure the confidentiality of patients' personal information, PEACE defines different access privileges for data requesters based on their roles, and then assigns different sets of attributes to data requesters. Using these different attribute sets, patient-centric access policies are proposed. PEACE can guarantee data integrity and confidentiality by adopting digital signature and pseudo-identity techniques. It embraces identity-based cryptography to securely aggregate remote patient data.

Recently, in Cholakoska et al. (2021), a differentially private federated learning method for anomaly detection in e-health networks was presented. Network traffic was protected using federated learning trained in conjunction with an anomaly detection system. This enabled real-time identification of traffic anomalies, while maintaining hospital collaboration and ensuring the security and confidentiality of local data.

3.5.2.3. *Security models for data storage and access*

Storing sensitive information on centralized servers raises the problem of data leakage. Data must therefore be adequately protected. In addition, client systems can be a source of security problems, that is, doctors' PCs, which are used both to manage patient data and to surf the Internet, for example.

Gajanayake et al. (2012) proposed a privacy-focused access control model to meet e-health requirements. The model was designed by combining three existing access control models (DAC, MAC and RBAC) into a new module that enables patients and healthcare professionals to determine and define access privileges. The module has been tested to demonstrate different policy settings and data access scenarios.

The Context-Aware Access Control Security Model (CARE) architecture is based on the data interoperability scenario and supports the fundamentals of healthcare system security, as well as the ability to provide precise access control (Altamimi and Mousa 2016). In CARE, policies are defined using the user interface module, which can be a website, or a mobile app. Patients and doctors together define policies to preserve patient privacy. All defined policies are then analyzed in their components (e.g. constraints) and stored in a centralized security database, which is represented using an extended RBAC model. ERBAC consists of roles, authorizations and users. Roles have been created for various professional functions, with authorizations for specific operations. Users are assigned specific roles and, through these role assignments, acquire authorizations to perform certain operations. Consolidating access control for many users into a single role entry makes it much easier to manage the whole system, and much more efficient to verify security policies. Three different types of contextual data have been taken into account in this model. Environmental context refers to location or time, personal information context refers to age, status or medical record, and finally, current medical data, such as heart rate or blood pressure.

Khan and Sakamura (2012) proposed a contextual approach to access control, based on the conventional models of DAC and RBAC. By adopting a holistic view of access control, it has effectively addressed these four constituent stages of identification, authentication, authorization and access decision. The eTRON (Entity and Economy TRON) architecture, which advocates the use of tamper-proof chips with mutual authentication and encrypted communication functions, is used for authentication and the implementation of DAC-based access control rights delegation. The RBAC model is used for the authorization and access decision stages.

3.6. Future security challenges for e-health

The main security challenges for e-health systems concern the use of blockchain, the cloud and compliance with the Health Information Technology for Economic and Clinical Health (HITECH) Act (Alenoghena et al. 2022).

Despite blockchain's suitability as a security solution for e-health systems, it remains an open ledger that is visible to everyone. This allows anyone to verify the accuracy of the blockchain. However, this can become a handicap, particularly in delicate contexts such as those linked to e-health system configuration. In this situation, further research may be needed to address privacy leakage issues, which may include developing information about the use of blending services. The use of multiple services can ensure anonymity by matching multiple input addresses to multiple output addresses, making it difficult to determine the origin of the information.

The use of cloud services for e-health raises issues of access control, privacy and compliance. Indeed, despite the multitude of cloud services, the storing of personal health data on public servers, patients lose the possibility of physical control over their data, which increases the risk of loss of confidentiality. Moreover, storing and accessing data in the cloud can be associated with integrity and confidentiality risks.

On another note, the HITECH Act and its regulations pose a number of challenges, including:

– the absence of strict access control measures, such as those enabling emergency access protocols, automatic disconnections and physical restrictions;

– inadequate audit systems associated with monitoring activities;

– the need to improve security mechanisms for data transmission over wireless channels;

– the need for authentication systems capable of implementing single identity, multi-factor authentication and role-based authorization.

Several proposals have used role-based access control to ensure the security of e-health systems. However, one avenue for further research would be the use of attribute-based access control (ABAC) architecture, which would offer greater flexibility and scalability for authentication and authorization.

3.7. Conclusion

The issues surrounding the SP implemented by a system are of capital importance to its life and development. This is why we felt it necessary to offer it a suitable framework in which to express, manipulate and implement it with the utmost care and risk reduction. More specifically, we considered that to achieve such a goal, it is necessary to be able to associate it with adequate specification and testing techniques.

Cybersecurity incidents affecting e-health services and infrastructures have a considerable impact. It is therefore essential to define a suitable SP focusing, among other things, on the availability, continuity and resilience of these systems and infrastructures. Issues such as data integrity, data protection and data confidentiality are, of course, always important when we talk about e-health.

In this chapter, we introduced the concept of executable SP, a concept used to facilitate the specification and testing of SPs. Then we listed the security services in e-health systems, as well as the security requirements in e-health systems. This led us to recall some health security policies for a few countries. Finally, we presented three levels of security models for e-health systems, namely during data collection, during transmission and during data storage.

3.8. References

Abassi, R. and Guemara El Fatmi, S. (2019). Security policies a formal environment for a test cases generation. In *Artificial Intelligence and Security Challenges in Emerging Networks*, Abassi, R. (ed.), IGI Global, Hershey, PA, 237–264.

Alenoghena, C.O., Onumanyi, A.J., Ohize, H.O., Adejo, A.O., Oligbi, M., Ali, S.I., Okoh, S.A. (2022). e-Health: A survey of architectures, developments in mHealth, security concerns and solutions. *International Journal of Environmental Research and Public Health*, 19(20), 13071.

Altamimi, A.M. (2016). Security and privacy issues in e-Healthcare systems: Towards trusted services. *International Journal of Advanced Computer Science and Applications*, 7(9), 229–236.

Barua, M., Liang, X., Lu, R., Shen, X. (2011). PEACE: An efficient and secure patient-centric access control scheme for e-Health care system. In *IEEE Conference on Computer Communications Workshops (INFOCOM WKSHPS)*, Shanghai.

Beltran, L., Cifuentes, Y., Ramirez, L. (2015). Analysis of security vulnerabilities for mobile health applications. *International Scholarly and Scientific Research & Innovation*, 9, 1067–1072.

Cholakoska, A., Pfitzner, B., Gjoreski, H., Rakovic, V., Arnrich, B., Kalendar, M. (2021). Differentially private federated learning for anomaly detection in e-health networks. In *Adjunct Proceedings of the 2021 ACM International Joint Conference on Pervasive and Ubiquitous Computing and Proceedings of the 2021 ACM International Symposium on Wearable Computers*.

Cristiá, M. and Rossi, G. (2021). Automated proof of Bell–LaPadula security properties. *Journal of Automated Reasoning*, 65(4), 463–478.

Eysenbach, G. (2001). What is e-health? *Journal of Medical Internet Research*, 3(2), E20.

Fraser, B. (1997). RFC 2196 site security handbook.

Gajanayake, R., Lane, B., Iannella, R., Sahama, T. (2012). Legal issues related to accountable-e-Health systems in Australia. *1st Australian e-Health Informatics and Security Conference (AeHIS)*, Perth.

Garcia-Morchon, O. and Wehrle, K. (2010). Efficient and context-aware access control for pervasive medical sensor networks. In *Pervasive Computing and Communications Workshops (PERCOM Workshops)*, Mannheim.

Ghafur, S., Kristensen, S., Honeyford, K., Martin, G., Darzi, A., Aylin, P. (2019). A retrospective impact analysis of the WannaCry cyberattack on the NHS. *npj Digit Med.*, 2, 98.

Kashmar, N., Adda, M., Atieh, M., Ibrahim, H. (2021). A review of access control meta-models. *Procedia Computer Science*, 184, 445–452.

Khan, M.F.F. and Sakamura, K. (2012). Context-aware access control for clinical information systems. In *International Conference on Innovations in Information Technology (IIT)*. IEEE, Abu Dhabi.

Ksibi, S., Jaidi, F., Bouhoula, A. (2022). A comprehensive study of security and cyber-security risk management within e-health systems: Synthesis, analysis and a novel quantified approach. *Mobile Networks and Applications*, 1–21.

Lin, X., Lu, R., Shen, X., Nemoto, Y., Kato, N. (2009). Sage: A strong privacy-preserving scheme against global eavesdropping for e-health systems. *IEEE Journal on Selected Areas in Communications*, 27(4), 365–378.

Oh, S., Young-Duk, S., Euijong, L., Young-Gab, K. (2021). A comprehensive survey on security and privacy for electronic health data. *International Journal of Environmental Research and Public Health*, 18(18), 9668.

OWASP Foundation (2022). OWASP [Online]. Available at: https://www.owasp.org/index.php/Main_Page.

Sawand, A., Djahel, S., Zhang, Z., Nait-Abdesselam, F. (2015). Toward energy-efficient and trustworthy e-Health monitoring system. *China Commun.*, 12, 46–65.

Security and Resilience in e-Health (2015). European Union agency for network and information security. Report.

Sterne, D.F. (1991). On the buzzword security policy. In *IEEE Symposium on Security and Privacy*.

Zriqat, I.A. and Altamimi, A.M. (2016). Security and privacy issues in e-healthcare systems: Towards trusted services. *International Journal of Advanced Computer Science and Applications (IJACSA)*, 7(9).

4

Adaptive, Dynamic, Decentralized Authorizations for e-Health

Tidiane SYLLA[1], Mohamed AYMEN CHALOUF[2], Léo MENDIBOURE[3] and Francine KRIEF[4]

[1] GEII, ISA, USTTB, Bamako, Mali

[2] IRISA, Université de Rennes, France

[3] ERENA, COSYS, Université Gustave Eiffel, Marne-la-Vallée, France

[4] LaBRI, Bordeaux INP, France

4.1. Introduction

The Internet of Things (IoT) is a major development in information and communication technologies, enabling any object equipped with sensors and/or actuators to be connected to the Internet to collect data and/or remotely action equipment (Li et al. 2015). The applications of IoT are numerous, improving users' daily lives. e-Health (or digital health) is an important technological branch propelled by the advances brought by the IoT in data collection and analysis. This is made possible by a wide range of connected devices (e.g. body-worn and non-body-worn) for collecting data on the health of users (patients).

e-Health is a constantly evolving field that uses information and communication technologies to improve healthcare and patients' quality of

e-Health Security Management,
coordinated by Omessaad HAMDI.

life. This interdisciplinary field mainly encompasses medical informatics, connected medical devices, mobile health applications, electronic health records, telemedicine, e-health and remote health services (Eysenbach 2001; Hamdi et al. 2014; Sylla 2021). e-Health applications are made up of a number of objects comprising sensors of all kinds, enabling information to be collected at any time, in any place, such as blood sugar levels, blood pressure, heart rate, sleep quality, physical activity and breathing levels.

e-Health facilitates the collection, storage, management and analysis of health data, which can help healthcare professionals make more informed decisions and deliver more effective, personalized healthcare (Ouattara et al. 2014). It also enables patients to access health information more easily, monitor their health status and communicate with their doctor and other healthcare professionals.

e-Health offers many advantages, but it also raises questions about the privacy and security of health data (Butpheng et al. 2020). It is therefore essential to research and propose robust security and confidentiality standards and solutions to protect patients' health information. Security in this field concerns data storage, communication and access. Managing access control to patient data is a crucial issue for security and privacy. In this chapter, we focus on authorization management in the context of e-health. Indeed, authorization management mechanisms can be used to control access to protected resources (e.g. medical sensors).

Several solutions have been proposed for access control and authorization management for IoT in general, and for e-health in particular (Arfaoui et al. 2018, 2019; Atlam et al. 2018; Chaudhry et al. 2020; Liu et al. 2020). Among them, we have the Authentication and Authorization for Constrained Environment (ACE) security framework. ACE has been proposed and standardized by the IETF (Internet Engineering Task Force) for authentication and authorization enabling access to resources identified by a URI and hosted on a constrained resource server (e-health devices: glucose meter, blood pressure monitor, etc.) (Seitz et al. 2018). However, few works consider context awareness in the implementation of these mechanisms (Chaudhry et al. 2020; Liu et al. 2020; Ameer et al. 2022). Yet the consideration of context awareness in authorization management is important, particularly in the e-health domain, by enabling the dynamism of e-health environments to be supported (Perera et al. 2014; Sylla et al.

2021a). For example, it is possible to grant contextual permissions to a doctor that allow them to consult a patient's health data while in hospital, but not to access the same data when the doctor is outside the hospital (e.g. from home). Another example would be to give the patient the option of assigning exceptional, context aware access rights to the data on their smart glucose meter to a nurse, in the event of a critical situation in hospital. Under normal circumstances (i.e. in a scheduled hospital situation), however, the nurse has only limited access to the glucometer's data.

Furthermore, existing solutions rely on centralized authorization servers. A central authorization server can represent a single point of failure and a performance bottleneck (Alphand et al. 2018). In addition, a central server can be compromised and expose the entire e-health system to attack. One possible solution is to base the authorization server on a distributed and implicitly trusted system such as blockchain in order to increase the security and robustness of authorization management (Alkurdi et al. 2018). Indeed, blockchain is a decentralized and trusted system, difficult to compromise (Sylla et al. 2021c).

In this chapter, we propose a dynamic, decentralized authorization management adaptation system for e-health. In section 4.2, we begin with a reminder of the fundamental concepts required to understand the rest of this chapter. In section 4.3, we present our proposed dynamic, decentralized context-aware authorization management system for e-health. Section 4.4 concludes the chapter.

4.2. Fundamental principles

In this section, we introduce the main concepts discussed in this chapter: e-health, context-aware security as a service, the blockchain distributed ledger and the ACE standard.

4.2.1. *Concept of e-health*

e-Health is not a new concept. In fact, the concept of e-health emerged at the end of the 1990s, at a time when new applications made possible by advances in the Internet (e-commerce, e-mail, e-government and e-banking) were making their appearance. The medical industry came up with the concept of e-health to keep up with the everything-electronic trend and promoted this concept, which was not yet accessible at the time. In the early 2000s, Eysenbach was one of the first to propose a scientific definition of e-health:

> e-Health is an emerging field at the intersection of medical informatics, public health and business, referring to health services and information provided or enhanced by the Internet and related technologies. (Eysenbach 2001, translated by authors)

As announced in the introduction to this chapter, the IoT has propelled e-health into a new dimension by enabling the various players to implement new applications enabled by the widespread availability of increasingly miniaturized sensors. e-Health applications are divided into several categories, such as:

– The assisted living environment: these applications aim to give the elderly more autonomy by providing them with the care and support they need. To achieve this, a multitude of sensors and cameras are installed in this environment (specialized home) (Chandra et al. 2019). These applications also make it possible to collect, organize and evaluate data on patients' activities, which can enable remote caregivers to respond quickly to emergencies.

– Connected health: also known as health IoT (HIoT), this covers all technologies, particularly IoT, that enable patients' health status to be monitored by collecting data from health sensors and other connected devices (connected glucose meter, connected blood pressure monitor, etc.) (Farahani et al. 2020). The data collected are rapidly analyzed by advanced algorithms in the cloud to diagnose and propose treatments to the patient under the supervision of a doctor.

– Electronic medical records: these enable healthcare professionals to store, consult and share medical information on patients (Oborn et al. 2011). The aim of sharing electronic medical records is to establish a reliable diagnosis and minimize medical errors. An electronic medical record contains data relating to the patient's medical history (admission, treatments, etc.), biomedical test results, prescribed medications, allergies, etc.

– Mobile health applications: these enable users to monitor their health, receive reminders to take their medication, track their diet and physical activity, etc. (Barton 2012; Boudreaux et al. 2014). Mobile health applications are subdivided into several subgroups: health education and awareness (provide information on disease prevention), patient monitoring, health point-of-care and clinician support, disease and epidemic surveillance, emergency medical response system, etc. (Boudreaux et al. 2014).

– Telemedicine: the practice of medicine at a distance, enabling patients to receive medical care at a distance using technologies such as videoconferencing, instant messaging or other information and communication technologies. This branch of e-health improves access to healthcare for patients living in areas where access to healthcare professionals is limited. It also makes it possible to respond to urgent care needs by providing remote assistance from specialized healthcare professionals.

In the remainder of this chapter, we will refer to e-health applications based on IoT technologies.

4.2.2. *Context-aware computing and security in the IoT*

4.2.2.1. *Context-aware computing in the IoT*

Intelligence, an important feature of the IoT, enables these applications to dynamically adapt their behavior to different user situations. Adaptation actions can be security-related or linked to the business logic of the application concerned. For example, they may involve opening the garage gate as the user approaches the house or choosing single-factor or two-factor authentication depending on location and the communication network used. Dynamic adaptation in the IoT is made possible by context-aware computing (Perera et al. 2014).

Context-aware computing in the IoT consists of using contextually relevant information to improve the services offered to the user through dynamic actions and decisions taken automatically, without requiring user intervention (Sylla et al. 2021a). Contextual information is collected from information sent by IoT devices, processed and analyzed by simple ontology or artificial intelligence algorithms to determine the user's context. This continuous process of detecting the user's context is known as the "context lifecycle", which is shown in Figure 4.1.

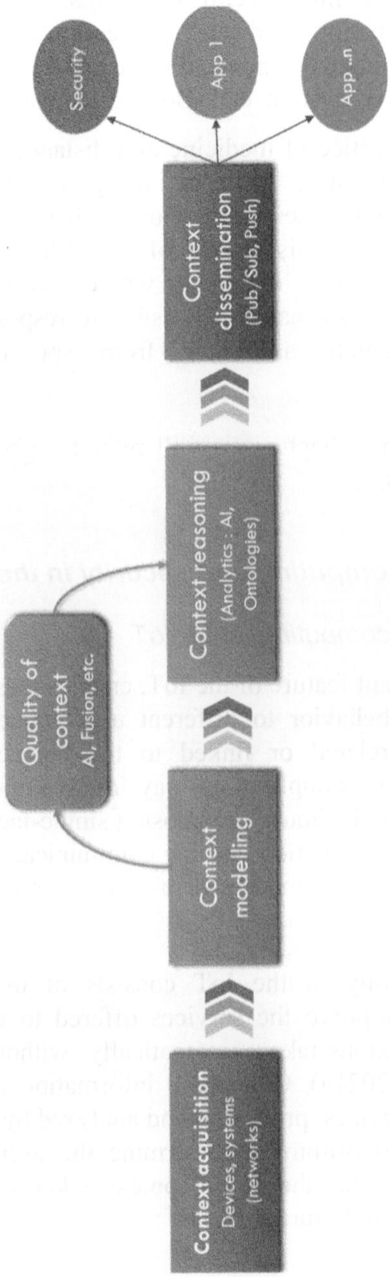

Figure 4.1. *Context lifecycle*

The context lifecycle can be modified to support the notion of context quality (CQ), which refers to the accuracy, relevance, reliability and usefulness of contextual information used to determine the user's context and make dynamic adaptive decisions or actions (Sylla et al. 2022). In this sense, CQ is important for ensuring the relevance of personalized services provided by context-aware systems. Incorrect, obsolete, imprecise or insufficient contextual information can lead the system to make dynamic adaptation decisions that are incorrect or useless for the user. To improve context quality, context-aware computing systems can use techniques such as machine learning, data analysis and fusion of different context sources to produce more accurate and relevant context information (Perera et al. 2014).

4.2.2.2. *Context-aware security in e-health*

Context-aware security in the IoT is a security approach that uses context-aware computing (taking into account factors such as the network environment and sensor data) and the risks associated with a given situation in the dynamic implementation of security and privacy mechanisms. This dynamic implementation takes place without user intervention (Sylla et al. 2021a).

This approach is particularly well suited to IoT applications in e-health, as the user context changes regularly in these applications. For example, at the hospital, the level of access authorizations to patient data when the patient is with a nurse at time t is not the same as when the patient is with their GP at time $t + 1$.

Context-aware security in new IoT e-health applications is essential to ensure and guarantee the confidentiality, integrity and availability of data, as well as to protect the devices themselves and the people who use them. Hospitals and organizations deploying IoT devices must ensure that they use adaptive security protocols that take into account contextual factors to reduce security risks. Indeed, each user situation can be associated with a set of threats that need to be addressed by alleviating/reinforcing the security and privacy mechanisms implemented.

Figure 4.2 illustrates this adaptive implementation.

Dynamic adaptation of the security level can be applied to mechanisms assuring the implementation of several security and privacy services in e-health (Alagar et al. 2018; Arfaoui et al. 2020). These include mechanisms implementing authentication, access control and authorization management, communications security and privacy protection (Neisse et al. 2015; Alagar et al. 2018; de Matos et al. 2018; Arfaoui et al. 2020; Sylla et al. 2021c).

Dynamic authorization management in e-health applications is particularly important, especially since in these environments, patients, because of their condition, may not be able to grant specific authorizations at the right time. Take, for example, an e-health application for monitoring a patient's cardiovascular health. Their heart rate data are transmitted from an ECG sensor to the cloud via a gateway. The doctor, who is at the hospital, can consult these data easily by connecting to the monitoring application. During the night, the application detects a tachycardia attack in the patient and sends an alert to the doctor. The doctor then tries to connect to the application to check the data. However, the system detects a new context (situation): the doctor is no longer at the hospital, it is 2 a.m. and he is using a 4G mobile network to connect. In this context, the security risks associated with access to the application, and therefore to patient data, are high. To avoid this, the system needs to implement an additional verification mechanism to ensure that it is indeed the doctor who is attempting to connect in this new context. This example shows that security risks can vary depending on the context, requiring the implementation of security services adapted to each situation. As a result, context-aware, dynamic authorization management can be an interesting security solution for e-health.

A context-aware security system exploits the different contexts provided by the context-aware management system (see section 4.2.2) to make dynamic adaptation decisions. Context-aware safety systems fall into two categories, depending on the architecture used. The first category involves an integrated system comprising both the context-aware manager and the modules for controlling and implementing security and privacy mechanisms. Solutions using this architecture are simpler and less complex to develop but can be cumbersome and inflexible in the dynamic implementation of dynamic adaptation decisions (Hayashi et al. 2013; Mowafi et al. 2014; Ashibani et al. 2017; de Matos et al. 2018; Arfaoui et al. 2020). In these architectures, each change of context implies the implementation of the whole set of mechanisms. In addition, solutions based on integrated architectures are difficult to maintain and are hard to manage with new network architectures (software-defined networks [SDN], network function of virtualization [NFV] and edge computing [EC]).

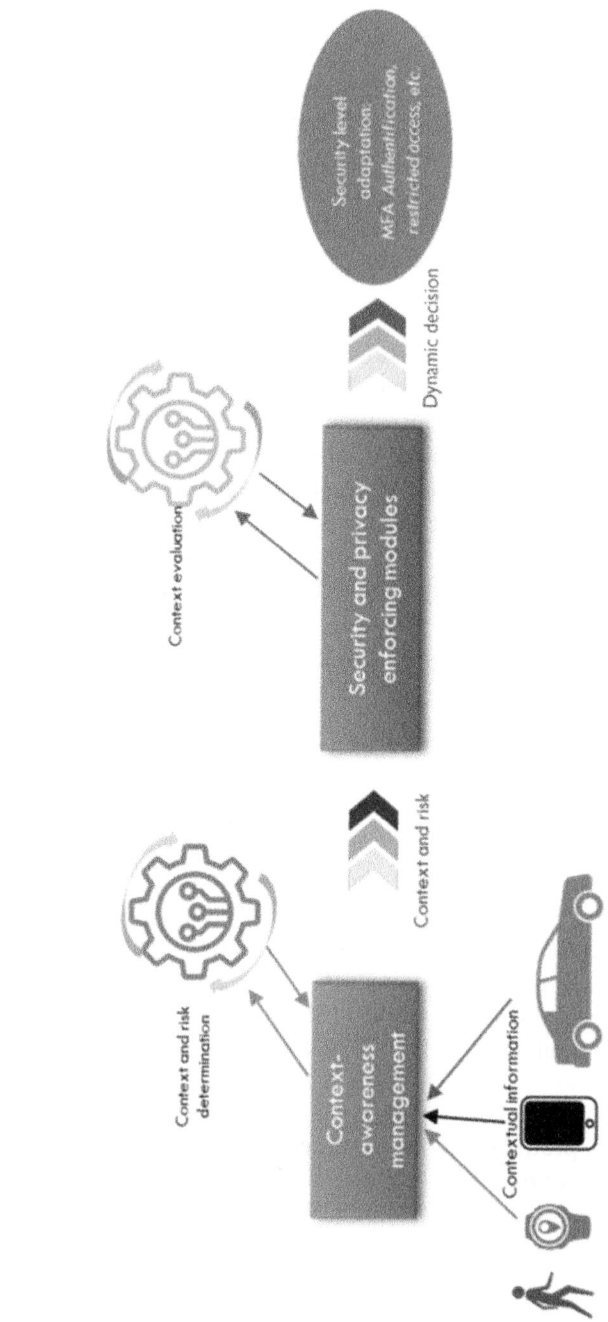

Figure 4.2. *Example of a context-aware security system*

The second category of context-aware security system is based on an architecture that separates the management of context awareness (knowledge plane) and the implementation of security and privacy mechanisms (security plane) (Sylla et al. 2020). In these architectures, intelligence, that is, the management of context awareness, is separated from the implementation of security decisions, and security and privacy solutions are more flexible and efficient for users. In addition, most solutions based on these architectures are designed according to the as-a-service approach, which facilitates their deployment in new network architectures (SDN, VNF and EC) (Boubendir et al. 2016, 2017).

Context awareness is the cornerstone of any context-aware security system. Context-aware management can be subject to a number of attacks (e.g. falsified contextual information) aimed at its proper functioning and impacting the dynamic implementation of adaptation decisions (Mahalle and Dhotre 2020; Sylla et al. 2021b). Therefore, the consideration of security and trust in the management of context awareness is a necessity. Secure context aware management in a trusted environment consists of implementing mechanisms to protect the context aware manager, secure contextual information exchanges and ensure that this information comes from secure context sources (e.g. trusted sensors) (Sylla 2021).

4.2.3. *Authentication and Authorization for Constrained Environments (ACE-OAuth)*

As mentioned in the introduction, the ACE-OAuth standard has been proposed to secure access to resources hosted by constrained devices, such as those mostly used by patients in e-health. Based on the OAuth 2.0 authentication protocol (Hardt 2012), it enables an authenticated third party holding an authorization token to access a protected resource via the Constrained Application Protocol (CoAP) (Shelby et al. 2014) and Message Queuing Telemetry Transport (MQTT) (Banks et al. 2019) application protocols.

The OAuth 2.0 protocol allows a user to authorize a third-party application to access its protected resources without revealing its credentials, using access tokens in JSON Web Token (JWT) format encoded with client information and authorized accesses. The JSON Object Signing and Encryption (JOSE) standard is used to secure the tokens (Jones 2015a, 2015b; Jones and Hildebrand 2015; Jones et al. 2015). Figure 4.3 shows the flow of exchanges between the various entities in the ACE-OAuth security framework. When a client obtains an access token and presents it to the resource server for access to the protected resource, the resource server may perform an additional introspection step (exchanges (4) and (5) in Figure 4.3) to ensure the validity of the access token.

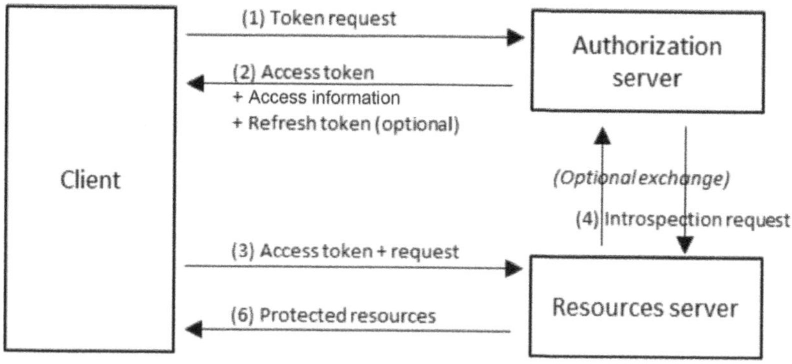

Figure 4.3. *ACE framework base protocol exchanges*

Figure 4.4 illustrates a secure e-health application using the ACE-OAuth security framework. This security framework defines several terms to designate the various entities. These include resource owner, resource server, authorization server and client. The resource owner is the legal owner of the resources to be protected, that is, the person whose data and/or devices are to be protected. Resource servers store and manage the protected resources (patient data). These are e-health devices such as smartwatches equipped with ECG sensors, connected glucometers and smart blood pressure monitors. In the ACE-OAuth framework, the authorization server acts as an intermediary between the resource owner, clients and resource servers. It is the party responsible for generating and validating the authentication and authorization information used by a resource server. Clients are the third parties who need access to the resources protected by the resource server. In e-health applications, this is usually an application with the following clients: a doctor, a nurse, a university hospital, etc.

Furthermore, to meet the requirements of constrained devices, such as those in the IoT, the IETF has proposed the Concise Binary Object Representation (CBOR) format (Bormann and Hoffman 2013). This format has been designed to facilitate the definition and transport of small messages for IoT applications. Geo-managed secure tokens according to the ACE-OAuth framework are represented in the CBOR Web Token (CWT) format (Jones et al. 2018). To secure CWT tokens, the CBOR Object Signing and Encryption (COSE) standard has been defined (Schaad 2016), as has JOSE for JWT tokens. The main difference between COSE and JOSE lies in the use of the CBOR format, concise and compact for the former, while the latter uses the JWT format.

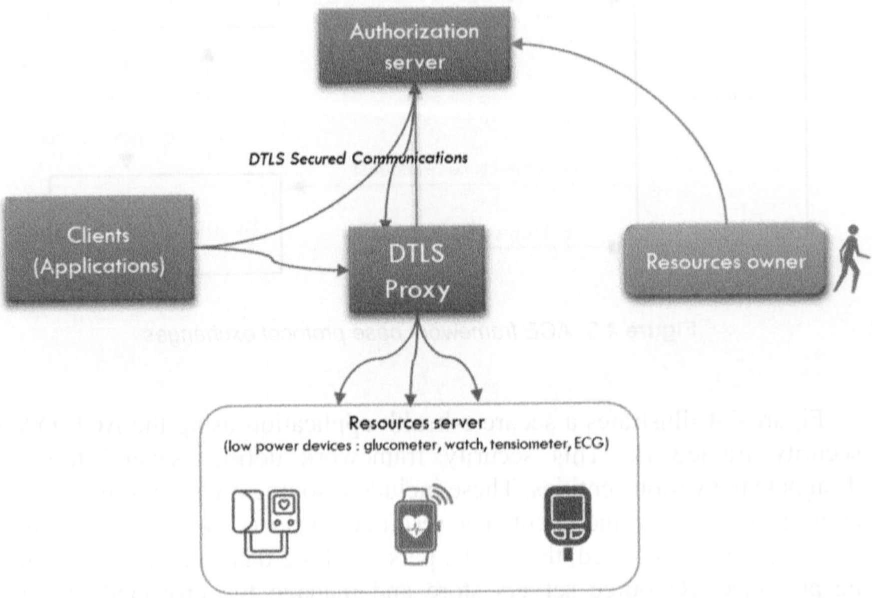

Figure 4.4. *ACE e-health security framework architecture*

Proof of possession (PoP) is a method used in ACE-OAuth to securely link issued access tokens to requests authorized by these tokens. To do this, a token is associated with a cryptographic key, enabling the client to prove that it possesses the secret key when accessing a resource. The resource server verifies that the key used by the client is the one associated with the token. The key can be symmetrical or asymmetrical. The PoP token contains four fields: the issuer identifier, the recipient identifier, the duration of the token's validity and a confirmation identifying the PoP key. Before processing the token, each client must identify itself to the token recipients.

4.2.4. *Blockchain*

Blockchain has represented a major breakthrough in the field of security. It is, in fact, the first approach to enable the secure transfer and storage of data without the need for a central control body (Mendiboure et al. 2020). Traditional systems for managing certificates, authorizations or currencies, for example, rely on the use of a trusted third party (bank, certification body, etc.). The latter is responsible for validating transfers and checking information.

With blockchain, network users can ensure the security of the data they wish to exchange/validate/store. This offers multiple advantages:

– Enhanced resilience: the removal of the network's single point of failure (central control body/trusted third party) increases the system's resilience, particularly in the face of denial of service (DoS) attacks.

– Extended deployment: the ability to verify/validate data locally, without the need to implement a global system involving internet connectivity, increases the number of scenarios in which the system can be deployed. Local networks of connected objects could thus be secured using blockchain.

– Scaling: the distributed architecture on which blockchain is based means that different nodes must offer the same services and be able to respond to customer requests. This seems to offer a clear advantage in a context of massive communication for connected objects.

The primary application of this technology is the storage, validation and securitization of data: cryptocurrencies, certificates and so on. Bitcoin was the first blockchain implementation to offer such a solution. However, some implementations of blockchain technology, such as Hyperledger or Ethereum, also include another important application: smart contracts (or chain codes). These are computer programs which, like transactions, are stored directly on the blockchain and are therefore not modifiable (Xie et al. 2019). These smart contracts can only be executed by authenticated and authorized blockchain users. They therefore offer a relevant (secure, distributed) approach to automating exchanges between users of the blockchain network. Some blockchain technologies also offer the possibility of interacting with these smart contracts via a graphical interface, known as a distributed application (DApp) (Christidis and Devetsikiotis 2016; Alkurdi et al. 2018). Transaction management, smart contracts and DApps are the three main applications of blockchain technology.

To offer these different applications, blockchain technology is based on an architecture composed of four main layers (see Figure 4.5):

– Applications layer: this layer corresponds to the smart contracts and DApps that are accessible within a given blockchain network. These applications represent a simplified solution for interacting with the data stored in the blockchain network. Each user of a blockchain network is likely to propose the deployment of new blockchain applications.

– Consensus layer: this is the main specificity of blockchain architecture. This trust layer corresponds to the mechanisms put in place to enable the secure exchange and storage of data without the need for a trusted third party. These are referred to as consensus algorithms, because of the need to reach agreement between the various blockchain nodes. Once this validation stage has been completed, the data (transactions, smart contracts) can be added to the "network data" layer. Numerous consensus algorithms have already been proposed, such as proof-of-work (PoW), proof-of-stake (PoS) and proof-of-authority (PoA).

– Communication layer: blockchain technology is based on direct exchanges between the various nodes in the network, creating a peer-to-peer network. This enables data to be distributed directly between the various nodes. Communications between these nodes are secured using cryptographic mechanisms (asymmetric keys).

– Data layer: blockchain uses a special data structure to store information. This is a chain of blocks. Each block corresponds to a set of data (transactions and smart contracts) that has been requested to be added to the blockchain network. These blocks are shared with all the nodes involved in the blockchain's operation. In addition to data, each block contains a hash from the previous block, the main mechanism for ensuring that the content of a block has not been altered, since there is a dependency between all the blocks stored and exchanged between blockchain nodes.

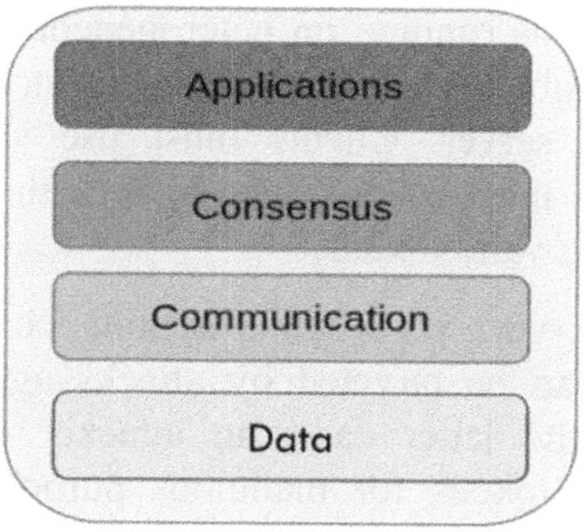

Figure 4.5. *A simplified view of blockchain architecture*

4.3. Proposal for dynamic, decentralized adaptation of e-health authorizations

In this section, we introduce the dynamic, decentralized context-aware authorization management system for e-health. To do so, we first present the threat model for this environment, which justifies the need for dynamic, decentralized authorization management. We will then introduce the proposed architecture, an extension of the ACE-OAuth framework. We will also pre-present the new context aware access token, the smart contract and the proposed new dynamic authorization flow.

4.3.1. *Threat model for the environment under consideration*

Accurate analysis of the threats and adversaries to which a system is exposed is an essential step in researching and designing a security solution for that system. This analysis is based on a threat model. The system under study is exposed to the following threats:

– Resource owner: may or may not be aware of the importance of security when interacting with an e-health application. It is crucial that they are aware of this importance, and that every interaction with the application is carried out via a secure communication channel. In addition, the user must be protected against any attempt at identity theft.

– Resource servers: these are constrained devices that can be targeted by multiple physical and network attacks. As such, they need to be secured to protect access tokens during processing and storage.

– Clients: applications running on heterogeneous, unconstrained devices (laptop, smartphone, tablet, etc.). These devices store access tokens issued by the authorization server. Clients must use secure communications. Created access tokens must be protected against theft, and clients must be prevented from forging valid tokens.

– Authorization server: it is an essential component of the ACE architecture, and it can be targeted by attacks against its availability or its compromise. In the latter case, an attacker could generate forged, non-legitimate access tokens for malicious purposes: data theft, identity theft, etc. The authorization server must therefore be robust, distributed and secure. It must therefore be robust, distributed and directly accessible by the resource owner.

– Context-aware security manager: its role, as described in section 4.3.2, is to determine and manage validity contexts according to the resource owner's preferences. It must be protected against attacks on its availability and functionality. Validity contexts must remain protected in the event of a commission by the context-aware security manager.

– Integrity of access tokens: it is necessary to guarantee that these tokens are protected against any attempt at illegitimate creation or modification. In addition, it is important to protect the validity contexts of these tokens against any attempt at illegitimate modification or creation (see section 4.3.2). A mechanism must be put in place to detect stolen or illegitimately created access tokens. In section 4.3.2, we present our proposed new architecture for decentralized, dynamic authorization management for e-health.

4.3.2. *Proposed architecture for dynamic, decentralized authorization management*

The proposed architecture replaces the access control system of the ACE framework with a new context-aware access control system based on context-aware access tokens. This new architecture includes several components of the ACE security framework, a Context Aware Security Manager (CASM) for e-health and the blockchain network, replacing the central ACE Edge computing infrastructure with a decentralized, distributed authorization server. The CASM comprises two sub-services: the knowledge management service (KMS) and the security mechanism deployment service (SDMS).

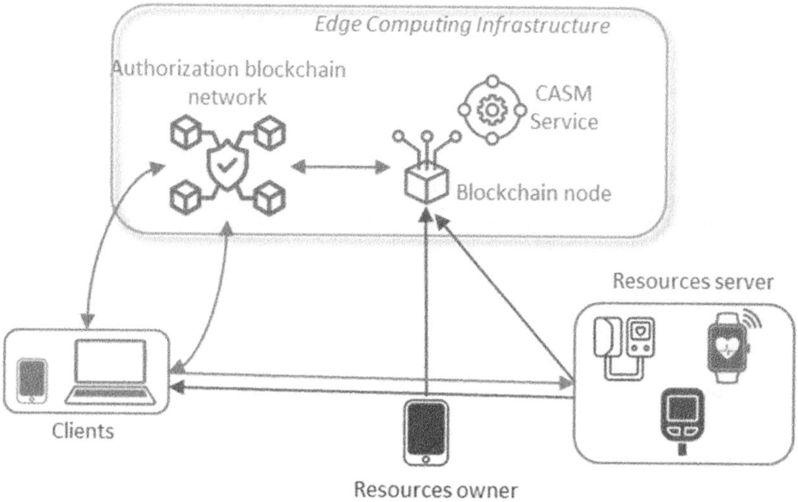

Figure 4.6. *Architecture for dynamic and decentralized adaptation of e-health authorizations architecture*

The KMS receives contextual information, analyzes it and provides the context to the SDMS. Depending on the context and risks, the SDMS activates the mechanisms required to secure the context. In this chapter, we focus solely on the deployment of the SDMS authentication and authorization module.

To stay close to the user and support mobility, the GSSC can be deployed in an edge computing (IB) infrastructure (Sabella et al. 2016; Sonmez et al. 2019). The orchestrator of an IB can launch and move the GSSC into IB servers at points of presence close to the patient (Sylla 2021). Figure 4.6 illustrates the new architecture for dynamic, decentralized authorization adaptation proposed for e-health.

The blockchain authorization network, made up of blockchain nodes, executes smart contracts responsible for generating context aware access tokens at the behest of the resource owner. Clients possessing valid context aware access tokens can access protected resources if and only if the patient's context so permits. Most constrained devices are generally able to perform lightweight asymmetric cryptographic operations and possess root certificates. They are also capable of contacting a certification authority. Devices unable to do so must use a proxy server to perform these operations (asymmetric cryptography). In this way, the devices concerned (resource servers) will be able to sign and verify the identities linked to the signatures via this proxy server. Table 4.1 summarizes the advantages of the proposed solution for authorization management in e-health.

Solutions	Authorization server	Server performance	Server security	Access token	Mobility
Classic authorization	Centralized	– Bottleneck – Difficult to scale up	Can be compromised	– Static access token – No revocation after issuance	No
Dynamic, decentralized adaptation	Distributed and decentralized (blockchain)	– No bottlenecks – Implicit scaling	Very difficult to compromise	– Context-aware, dynamic access token – On-the-fly revocation at user's request	Possible with edge computing

Table 4.1. *Comparison of the proposed solution with a conventional solution*

4.3.2.1. *Proposed new authorization flow*

Implementation of the proposed authorization flow assumes that the patient, or resource owner, is initially registered in the authorization blockchain network, and that their e-health devices are also registered and configured in the CASM service. These registration steps are not covered in this chapter.

The authorization flow of the proposed decentralized architecture, illustrated in Figure 4.7, begins with the definition of patient preferences (Step 1a). The patient performs this operation via an interface in the CSSM. The patient can define, according to their needs, which e-health applications can obtain adaptive access authorization, which resources they can access, and under which conditions/situations they can obtain these authorizations. The CSSM sends the data required to generate context aware access tokens to the smart contract (Step 1b). Only the patient can authorize the generation of context aware access tokens. However, they can add an additional person who can authorize token generation in the event of an emergency.

When a request is received, the smart contract executes the "Generate Context Access Token" function to dynamically generate the secure context access token for a user-specified client (step 2).

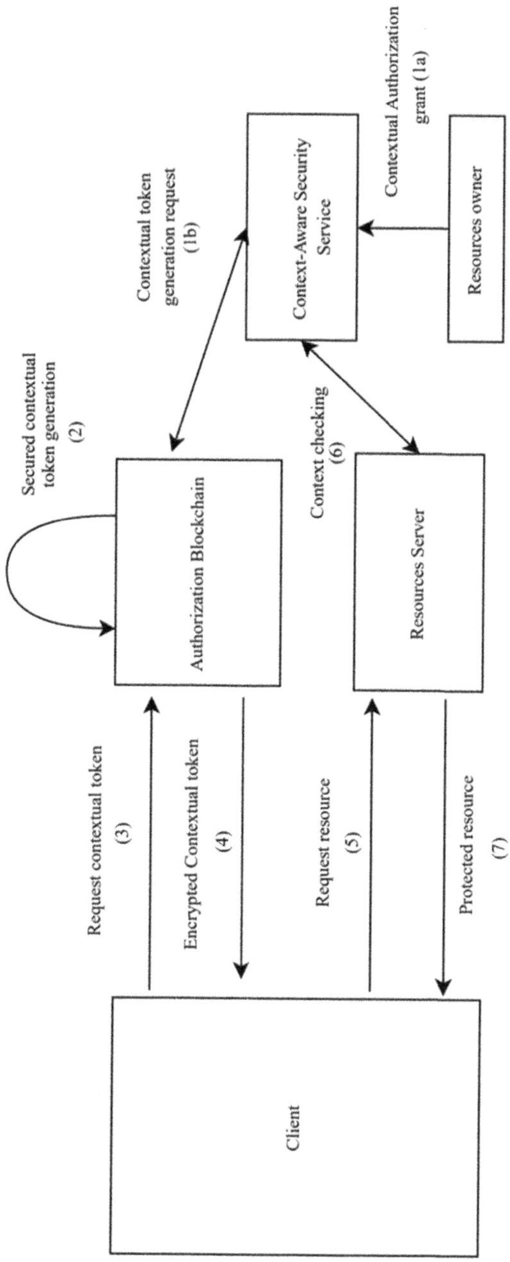

Figure 4.7. *Proposed new authorization flow for dynamic, decentralized adaptation of e-health authorizations*

The "Generate context aware access token" function of the smart contract is similar to the token endpoint of the classic authorization server in the ACE-OAuth security framework. The context-aware authorization tokens proposed here have the basic format of the ACE-OAuth framework's authorization tokens, with the addition of extra fields for managing dynamic adaptation to the user's context.

In addition, each context aware access token is linked to an identifier called the validity context identifier. Validity contexts (possible situations) are managed by the CSSM. The validity context is used by the smart contract to check the validity of the context access token with the CASM.

When a token is issued, its validity period is determined according to its specific use. A concrete example is OAuth 2.0, where the validity period of access tokens can be set between 1 and 60 days, depending on requirements. On the other hand, each context aware access token contains a digital fingerprint of its validity context identifier. Similar to the ACE-OAuth framework, the proposed context access token also contains digital fingerprints of the client's and resource server's public key, as well as two digital signatures generated by the smart contract to the client and resource server (in this case, the e-health device). This protects the context-aware access token against unauthorized modification. It also enables token recipients to verify that the tokens have been generated by the authorization blockchain's smart contract.

To access protected resources, the customer must perform several steps. First, he or she must contact the user's smart contract by calling the smart contract's "Request a context aware access token" function (Step 3). The smart contract signs the secure context aware access token and hands it to the customer (Step 4). When the customer receives the token, he or she must verify the signature on the smart contract before it can be decrypted. Once the contextual authorization token has been obtained, the client can present it to the resource server (Step 5). Depending on the situation there may be two possible scenarios: either the device can communicate directly with the client (Case 1), or the resource server is highly constrained and must go through a proxy server to communicate with the client (Case 2).

The second case requires additional processing at the proxy server level and will not be discussed in this chapter. The rest of the flow describes the first case, that is the resource server can communicate directly with the client. The process consists of several steps designed to guarantee secure access to protected resources. First, the resource server must authenticate the client by verifying the smart contract and client signatures. Next, it checks the lifetime of the token to ensure that it is still valid. If so, the resource server checks the client's proof of possession of the token. Verifying proof of possession is an important step in ensuring that the client is who they claim to be. If proof of possession is established, the resource server uses the token's context identifier to check whether the user's context matches that of the access token. This check is carried out by the department responsible for context-aware security.

Context checking is important because it takes into account the dynamic nature of the user's context at the time of the access request. Indeed, the patient's context at time t may be different from that at time $t + 1$. For example, in the case of a nurse, the context check can determine whether access to the client application (the nurse's application) should be authorized or rejected, depending on the critical situation at the hospital. If the token's context is valid, the client can access the protected resources. Finally, the security of communication between the client and the resource server, as well as the security of the token, are important aspects that are discussed in section 4.3.2.2. These aspects are crucial for guaranteeing the confidentiality, integrity and authenticity of data exchanged between the client and the resource server.

4.3.2.2. *Security of context aware access tokens and exchanges*

In this section, we describe the security of the context aware access tokens proposed in this chapter, as well as the security of their exchange in (authorization flow).

4.3.2.2.1. Context aware access token security

Context aware access token security is the cornerstone of a dynamic, decentralized authorization system for e-health. Context aware access tokens must be protected against forgery, unauthorized modification and interception. The blockchain authorization network's smart contract generates context-aware access tokens in JWT format, enabling clients to access protected resources on the resource server. Before the client can pre-submit the token to the resource server, it must be decrypted and put into COSE format to be compatible with the ACE-OAuth security framework.

When generating a context-aware access token, the smart contract encrypts the JWT object using the recipient client's public key and signs the token using the Elliptic Curve Digital Signature Algorithm (ECDSA) (Johnson et al. 2001; Bauer et al. 2016). Only clients previously authorized by the smart contract can call the context aware access token request function. On receipt of the context aware access token, the client verifies the signature of the smart contract and proceeds to decrypt the token. On successful decryption, the client encodes the token in COSE format, adds its own signature and transfers the COSE object containing the encrypted context access token to the resource server. To do this, the client establishes secure communication with the resource server to ensure the confidentiality and integrity of the data exchanged.

When the resource server receives the encrypted context access token under the COSE for-mat, it first decrypts it using its own private key. It then verifies the smart contract signature using the corresponding public key. If the signature is valid, the resource server can then proceed to verify the PoP included in the token to ensure that the client is who they claim to be. If the smart contract signature does not match, then the resource server must reject the token as invalid. Similarly, if the token's PoP is not established, then the resource server must also reject the token.

In summary, the process of generating and transmitting context aware access tokens over the authorization blockchain network is secure and efficient, offering a high level of protection for protected resources and user data. In addition, verification of the smart contract signature is an important step in the process of checking the validity of a context aware access token and must be carried out by the resource server to guarantee secure access to protected resources.

4.3.2.2.2. Secure exchanges

Communications between the resource owner and the CSSM, as well as between the customer and the blockchain, will take place via secure communication channels using the Transport Layer Security (TLS) protocol. This choice is justified by the adequate capabilities of the devices running the resource owners' and customers' applications, such as smartphones, tablets or laptops.

In addition, it is crucial to guarantee the security of exchanges between the client and the resource server, especially when these pass over unsecured underlying networks. In this case, securing the payload at the application layer can be an interesting alternative to securing the transport layer. In the context of ACE-OAuth security, the use of Ephemeral Diffie-Hellman Over

COSE (EDHOC) (Selander et al. 2019) is proposed for secure token exchanges between client and resource server. EDHOC makes it possible to establish a secure communication channel suitable for constrained devices by using ephemeral keys that are created only for the duration of the communication.

The use of EDHOC also has the advantage of requiring few resources, which is particularly important in the context of IoT, where devices can be limited in terms of computing capacity and energy. For exchanges between client and resource server over unsecured underlying networks, the use of EDHOC is an interesting alternative for guaranteeing the security of COSE token exchanges (Selander et al. 2019).

4.3.2.3. *Revoking context aware access tokens*

The proposed system for dynamic, decentralized authorization adaptation as a service has the advantage of enabling dynamic revocation of authorization tokens or entities authorized by the user. In fact, this system enables the dynamic revocation of context aware access tokens, as well as clients.

4.3.2.3.1. Revoking a content access token

When the resource owner wishes to revoke a context aware access token assigned to a customer, it can do so using a revocation functionality provided by the smart contract. In this case, the patient sends a token revocation request to the CSSM, which uses this functionality to revoke the client's context aware access token.

Revoking a context aware access token may be necessary for a variety of reasons, such as when the user's context changes, as in the example given. In this case, the exceptional rights granted to the nurse are no longer required and must be withdrawn. Token revocation may also be necessary in the event of a security breach or fraud.

The revocation functionality provided by the smart contract can be implemented in different ways, depending on the specific needs of the application or system. It is important to note that the revocation of a context-aware access token must be carried out appropriately to guarantee the security of aware resources and data.

4.3.2.3.2. Revoking a client

As with adding a client, revoking a client is done easily and dynamically according to the user's preferences as defined in the CSSM. The user can revoke a client at any time. To do this, the CSSM calls up the "Revoke customer" function of the smart contract. The smart contract adds the customer to the list of revoked customers. Once this has been done, the context aware access tokens generated for this customer are invalidated.

4.4. Conclusion

IoT applications for improving healthcare will continue to develop, with advances in biomedical electronics, the growing miniaturization of sensors and the integration of IoT and artificial intelligence into new network architectures, such as 5G and edge computing. This development will have to be accompanied by research and the proposal of innovative security solutions tailored to the users of these applications.

In this chapter, we focused on adaptive, decentralized authorization management as a service for e-health. We began by presenting the requirements for access control and authorization management in IoT-based e-health applications. This enabled us to propose a new system for dynamic, decentralized authorization management as a service for IoT-based e-health apps.

The proposed solution is based on ACE-OAuth, standardized by the IETF, to ensure authentication and authorization management in constrained environments. It adds context awareness to the ACE-OAuth security framework to overcome the limitations of the traditional approach and replaces the central ACE-OAuth authorization server with blockchain for a more resilient system. Design as a service allows the system proposed as an urbanized service (Pujolle 2020) to be placed as a service in edge computing infrastructures for greater proximity to users and also to support user mobility. In addition, we presented in detail the proposed system, the interactions of the various adaptive and dynamic authorization management processes.

4.5. References

Alagar, V., Alsaig, A., Ormandjiva, O., Wan, K. (2018). Context-based security and privacy for healthcare IoT. In *2018 IEEE International Conference on Smart Internet of Things (SmartIoT)*. Xi'an. doi: 10.1109/SmartIoT.2018.00-14.

Alkurdi, F., Elgendi, I., Munasinghe, K.S., Sharma, D., Jamalipour, A. (2018). Blockchain in IoT security: A survey. In *28th International Telecommunication Networks and Applications Conference (ITNAC)*. IEEE, Sydney. doi: 10.1109/ATNAC.2018.8615409.

Alphand, O., Amoretti, M., Claeys, T., Dall'Asta, S., Duda, A., Ferrari, G., Rousseau, F., Tourancheau, B., Veltri, L., Zanichelli, F. (2018). IoT chain: A blockchain security architecture for the Internet of Things. In *2018 IEEE Wireless Communications and Networking Conference (WCNC)*, Barcelona. doi: 10.1109/WCNC.2018.8377385.

Ameer, S., Benson, J., Sandhu, R. (2022). An attribute-based approach toward a secured smart-home IoT access control and a comparison with a role-based approach. information. *Information*, 13(2), 60. doi: 10.3390/info13020060.

Arfaoui, A., Kribeche, A., Boudia, O.R.M., Ben Letaifa, A., Senouci, S.M., Hamdi, M. (2018). Context-aware authorization and anonymous authentication in wireless body area networks. In *2018 IEEE International Conference on Communications (ICC 2018)*, Kansas City. doi: 10.1109/ICC.2018.8422397.

Arfaoui, A., Cherkaoui, S., Kribeche, A., Senouci, S.M., Hamdi, M. (2019). Context-aware adaptive authentication and authorization in Internet of Things. In *2019 IEEE International Conference on Communications (ICC)*, Shanghai. doi: 10.1109/ICC.2019.8761830.

Arfaoui, A., Boudia, O.R.M., Kribeche, A., Senouci, S.M., Hamdi, M. (2020). Context-aware access control and anonymous authentication in WBAN. *Comput. Secur.*, 88, 101496. doi: 10.1016/j.cose.2019.03.017.

Ashibani, Y., Kauling, D., Mahmoud, Q.H. (2017). A context-aware authentication service for smart homes. In *14th IEEE Annual Consumer Communications & Networking Conference (CCNC)*, Las Vegas. doi: 10.1109/CCNC.2017.7983179.

Atlam, F., Alenezi, H., Khalid Hussein, A., Wills, G. (2018). Validation of an adaptive risk-based access control model for the Internet of Things. *Int. J. Comput. Netw. Inf. Secur.*, 10, 26–35. doi: 10.5815/ijcnis.2018.01.04.

Banks, A., Briggs, E., Borgendale, K., Gupta, R. (2019). MQTT Version 5.0 OASIS standard [Online]. Available at: https://docs.oasis-open.org/mqtt/mqtt/v5.0/os/mqtt-v5.0-os.html [Accessed 18 March 2020].

Barton, A.J. (2012). The regulation of mobile health applications. *BMC Med.*, 10, 46. doi: 10.1186/1741-7015-10-46.

Bauer, J., Staudemeyer, R.C., Pöhls, H.C., Fragkiadakis, A. (2016). ECDSA on things: IoT integrity protection in practise. In *Information and Communications Security*, Lam, K.Y., Chi, C.H., Qing, S. (eds). Springer, Berlin. doi: 10.1007/978-3-319-50011-9_1.

Bormann, C. and Hoffman, P. (2013). RFC 7049: Concise Binary Object Representation (CBOR). Standard, Internet Engineering Task Force (IETF). doi: 10.17487/rfc7049.

Boubendir, A., Bertin, E., Simoni, N. (2016). NaaS architecture through SDN-enabled NFV: Network openness towards web communication service providers. In *NOMS 2016 – 2016 IEEE/IFIP Network Operations and Management Symposium*. IEEE, Istanbul. doi: 10.1109/NOMS.2016.7502885.

Boubendir, A., Bertin, E., Simoni, N. (2017). A VNF-as-a-service design through micro-services disassembling the IMS. In *2017 20th Conference on Innovations in Clouds, Internet and Networks (ICIN)*. IEEE, Paris. doi: 10.1109/ICIN.2017.7899412.

Boudreaux, E.D., Waring, M.E., Hayes, R.B., Sadasivam, R.S., Mullen, S., Pagoto, S. (2014). Evaluating and selecting mobile health apps: Strategies for healthcare providers and healthcare organizations. *Transl. Behav. Med.*, 4, 363–371. doi: 10.1007/s13142-014-0293-9.

Butpheng, C., Yeh, K.H., Xiong, H. (2020). Security and privacy in IoT-cloud-based e-health system – A comprehensive review. *Symmetry*, 12, 1191. doi: 10.3390/sym12071191.

Chandra, I., Sivakumar, N., Gokulnath, C.B., Parthasarathy, P. (2019). IoT based fall detection and ambient assisted system for the elderly. *Clust. Comput.*, 22, 2517–2525. doi: 10.1007/s10586-018-2329-2.

Chaudhry, S.A., Yahya, K., Al-Turjman, F., Yang, M.-H. (2020). A secure and reliable device access control scheme for IoT based sensor cloud systems. *IEEE Access*, 8, 139244–139254. doi: 10.1109/ACCESS.2020.3012121.

Christidis, K. and Devetsikiotis, M. (2016). Blockchains and smart contracts for the Internet of Things. *IEEE Access*, 4, 2292–2303. doi: 10.1109/ACCESS.2016.2566339.

Eysenbach, G. (2001). What is e-health? *J. Med. Internet Res.*, 3. doi: 10.2196/jmir.3.2.e20.

Farahani, B., Firouzi, F., Chakrabarty, K. (2020). Healthcare IoT. In *Intelligent Internet of Things*, Firouzi, F., Chakrabarty, K., Nassif, S. (eds). Springer, Berlin, 515–545. doi: 10.1007/978-3-030-30367-9_11.

Hamdi, O., Chalouf, M.A., Ouattara, D., Krief, F. (2014). e-Health: Survey on research projects, comparative study of telemonitoring architectures and main issues. *J. Netw. Comput. Appl.*, 46, 100–112. doi: 10.1016/j.jnca.2014.07.026.

Hardt, D. (2012). RFC 6749: The OAuth 2.0 Authorization Framework. doi: 10.17487/rfc6749.

Hayashi, E., Das, S., Amini, S., Hong, J., Oakley, I. (2013). CASA: Context-aware scalable authentication. In *Proceedings of the Ninth Symposium on Usable Privacy and Security – SOUPS '13.* ACM Press, Newcastle. doi: 10.1145/2501604.2501607.

Johnson, D., Menezes, A., Vanstone, S. (2001). The elliptic curve digital signature algorithm (ECDSA). *Int. J. Inf. Secur.*, 1, 36–63. doi: 10.1007/s102070100002.

Jones, M. (2015a). RFC 7517: JSON Web Key (JWK). Report, Internet Engineering Task Force (IETF).

Jones, M. (2015b). RFC 7518: JSON Web Algorithms (JWA). Report, Internet Engineering Task Force (IETF).

Jones, M. and Hildebrand, J. (2015). RFC 7516: JSON Web Encryption (JWE). Report, Internet Engineering Task Force (IETF).

Jones, M., Bradley, J., Sakimura, N. (2015). RFC 7515: JSON Web Signature (JWS). Report, Internet Engineering Task Force (IETF).

Jones, M., Erdtman, S., Tschofenig, H. (2018). CBOR Web Token (CWT) [Online]. Available at: https://datatracker.ietf.org/doc/html/rfc8392 [Accessed 18 January 2021].

Li, S., Xu, L.D., Zhao, S. (2015). The internet of things: A survey. *Inf. Syst. Front.*, 17, 243–259. doi: 10.1007/s10796-014-9492-7.

Liu, H., Han, D., Li, D. (2020). Fabric-iot: A blockchain-based access control system in IoT. *IEEE Access*, 8, 18207–18218. doi: 10.1109/ACCESS.2020.2968492.

Mahalle, P.N. and Dhotre, P.S. (eds) (2020). Security issues in context-aware systems. In *Context-Aware Pervasive Systems and Applications*. Springer, Berlin. doi: 10.1007/978-981-32-9952-87.

de Matos, E., Tiburski, R.T., Amaral, L.A., Hessel, F. (2018). Providing context-aware security for IoT environments through context sharing feature. In *17th IEEE International Conference on Trust, Security and Privacy in Computing and Communications/12th IEEE International Conference on Big Data Science and Engineering (TrustCom/BigDataSE)*, New York. doi: 10.1109/TrustCom/BigDataSE.2018.00257.

Mendiboure, L., Chalouf, M.A., Krief, F. (2020). A scalable blockchain-based approach for authentication and access control in software defined vehicular networks. In *29th International Conference on Computer Communications and Networks (ICCCN)*. IEEE, Honolulu. doi: 10.1109/ICCCN49398.2020.9209661.

Mowafi, Y., Abou-Tair, D., Al-Aqarbeh, T., Abilov, M., Dmitriyev, V., Gomez, J.M. (2014). A context-aware adaptive security framework for mobile applications. In *Proceedings of the 3rd International Conference on Context-Aware Systems and Applications*. ICST, Dubai. doi: 10.4108/icst.iccasa.2014.257495.

Neisse, R., Steri, G., Fovino, I.N., Baldini, G. (2015). SecKit: A model-based security toolkit for the Internet of Things. *Comput. Secur.*, 54, 60–76. doi: 10.1016/j.cose.2015.06.002.

Oborn, E., Barrett, M., Davidson, E. (2011). Unity in diversity: Electronic patient record use in multidisciplinary practice. *Inf. Syst. Res.*, 22, 547–564. doi: 10.1287/isre.1110.0372.

Ouattara, D., Chalouf, M.A., Hamdi, O., Krief, F. (2014). Multimedia content delivery for remote patient monitoring using cognitive radio networks. In *2014 Wireless Telecommunications Symposium (WTS)*. IEEE, Washington. doi: 10.1109/WTS.2014.6835003.

Perera, C., Zaslavsky, A., Christen, P., Georgakopoulos, D. (2014). Context aware computing for the Internet of Things: A survey. *IEEE Commun. Surv. Tutor.*, 16, 414–454. doi: 10.1109/SURV.2013.042313.00197.

Pujolle, G. (2020). *Software Networks: Virtualization, SDN, 5G and Security*, 2nd edition. ISTE Ltd, London, and John Wiley & Sons, New York.

Sabella, D., Vaillant, A., Kuure, P., Rauschenbach, U., Giust, F. (2016). Mobile-edge computing architecture: The role of MEC in the Internet of Things. *IEEE Consum. Electron. Mag.*, 5, 84–91.

Schaad, J. (2016). CBOR object signing and encryption (COSE) [Online]. Available at: https://cose-wg.github.io/cose-spec/ [Accessed 20 June 2019].

Seitz, L., Selander, G., Wahlstroem, E., Erdtman, S., Tschofenig, H. (2018). Authentication and authorization for constrained environments (ACE) using the OAuth 2.0 Framework (ACE-OAuth) (ACE Working Group Internet-Draft). Report, Internet Engineering Task Force (IETF).

Selander, G., Mattsson, J., Palombini, F., Seitz, L. (2019). RFC 8613: Object Security for Constrained RESTful Environments (OSCORE). Standard, Internet Engineering Task Force (IETF). doi: 10.17487/RFC8613.

Shelby, Z., Hartke, K., Bormann, C. (2014). RFC 7252: The Constrained Application Protocol (CoAP). Standard, Internet Engineering Task Force (IETF). doi: 10.17487/rfc7252.

Sonmez, C., Ozgovde, A., Ersoy, C. (2019). Fuzzy workload orchestration for edge computing. *IEEE Trans. Netw. Serv. Manag.*, 16, 769–782. doi: 10.1109/TNSM.2019.2901346.

Sylla, T. (2021). Sécurité et vie privée centrées sur l'utilisateur dans l'IoT. PhD Thesis, Université de Bordeaux, Bordeaux.

Sylla, T., Chalouf, M.A., Krief, F., Samaké, K. (2020). Towards a context-aware security and privacy as a service in the Internet of Things. In *Information Security Theory and Practice*, Laurent, M. and Giannetsos, T. (eds). Springer, Berlin. doi: 10.1007/978-3-030-41702-4_15.

Sylla, T., Chalouf, M.A., Francine, K., Samake, K. (2021a). Context-aware security in the Internet of Things: A survey. *Int. J. Auton. Adapt. Commun. Syst.*, 14, 1. doi: 10.1504/IJAACS.2021.10035751.

Sylla, T., Chalouf, M.A., Krief, F., Samaké, K. (2021b). SETUCOM: Secure and trust-worthy context management for context-aware security and privacy in the Internet of Things. *Secur. Commun. Netw.*, 1–24. doi: 10.1155/2021/6632747.

Sylla, T., Mendiboure, L., Chalouf, M.A., Krief, F. (2021c). Blockchain-based context-aware authorization management as a service in IoT. *Sensors*, 21, 7656. doi: 10.3390/s21227656.

Sylla, T., Chalouf, M.A., Krief, F. (2022). Adapting the security level of IoT applications. In *Intelligent Security Management and Control in the IoT*, Chalouf, M.A. (ed.). ISTE Ltd, London, and John Wiley & Sons, New York.

Xie, J., Tang, H., Huang, T., Yu, F.R., Xie, R., Liu, J., Liu, Y. (2019). A survey of block-chain technology applied to smart cities: Research issues and challenges. *IEEE Commun. Surv. Tutor.* doi: 10.1109/COMST.2019.2899617.

5

Applying Blockchain to e-Health

Cyrine LAHSINI[1], Faiza HAMDI[2] and Omessaad HAMDI[3]

[1] *Sys'Com-ENIT, Université de Tunis El Manar, Tunis, Tunisia*

[2] *Jeddah University, Saudi Arabia*

[3] *IEEE, Rennes, France*

5.1. Introduction

Blockchain technology, proposed by Satoshi Nakamoto in his article (Nakamoto and Bitcoin 2008), was born in 2008, as the basis of bitcoin technology, a peer-to-peer electronic payment system. Since then, this technology has revolutionized the industry, becoming one of the most significant advances of our time. It is based on the principles of trust and transparency and offers a decentralized environment.

Blockchain technology has attracted a great deal of interest from companies and researchers looking to apply it in various sectors, particularly healthcare. Data management within connected healthcare systems, also known as e-health systems, is a crucial process aimed at storing and analyzing medical information in order to improve disease treatment and propose more effective therapies. Efficient data management also enables disease prevention.

Today, e-health systems involve many different entities, such as healthcare professionals, hospitals, pharmacies, insurance companies, researchers and patients. However, the diversity and dispersion of these

e-Health Security Management,
coordinated by Omessaad HAMDI. © ISTE Ltd 2024.

actors poses challenges for the healthcare sector, particularly in terms of data security.

e-Health systems face numerous security issues. Blockchain technology may present a promising solution for securing patient information and ensuring the confidentiality and interoperability of healthcare data. Thanks to its decentralized nature and inalterability, this technology could ensure the integrity of healthcare data. Numerous blockchain implementations are available for several e-health applications (Xie et al. 2021).

In this chapter, we present a brief description of blockchain technology. This is followed by a description of the healthcare sector, exploring its particularities and the contribution of using blockchain technology in the medical sector.

5.2. Blockchain technology

5.2.1. *Blockchain fundamentals*

Blockchain technology is an emerging technology full of promise for various sectors of the continuously connected society. The use of blockchain technology has offered various solutions to address different aspects of security, while ensuring decentralized access to the end user (the patient in the healthcare sector). It is considered to be "a network of computers that all have the same transaction history" (Krishnan et al. 2020, translated by authors).

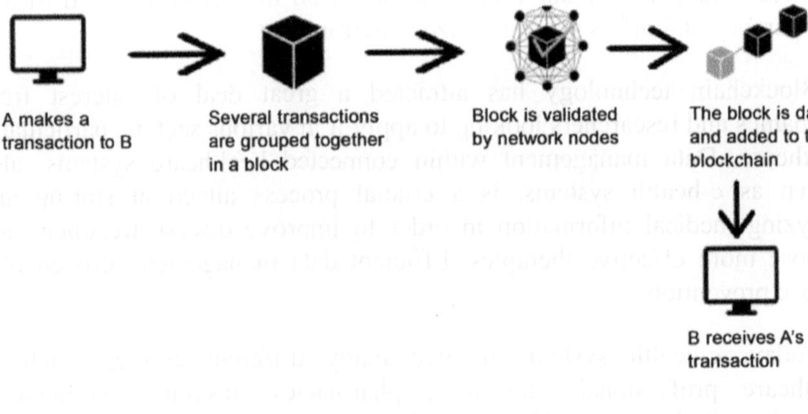

Figure 5.1. *How blockchain technology works (Dellys 2020)*

A blockchain is used to organize transactions in the form of blocks, which are verified by several participants who all store a copy, and then work together in the process of validating and certifying the transactions. The operation of a blockchain is illustrated in Figure 5.1 (Dellys 2020).

The blockchain is based on a decentralized system made up of six layers:

– Data layer: this layer generally presents the data object ready in a browser for integration with another party. It is based on string structure, data block, timestamp, Merkle tree and cryptography.

– Network layer: it includes router packets and switches, creating logical associations between remote stations and providing services. The basic components of this layer are packet sequencing, the broadcast mechanism and the verification mechanism.

– Consensus layer: its purpose is to verify that the actual state of the system is maintained.

– Contract layer: it includes smart contracts, script coding, consensus protocols and incentive mechanisms.

– Service layer: this is a conceptual layer within a network service provider architecture, with Ethereum and Hyperledger as core components.

– Application layer: this layer aims to put blockchain-based activities into practice.

Blockchain is primarily based on two major open-source platforms:

– Ethereum: it is an open platform that enables the creation, execution and use of decentralized applications. Ethereum enables the development of applications that facilitate and automate direct exchanges and interactions between network members. Ethereum enables the creation and development of a system without any intermediary authority. Improved application security, the ability to interact efficiently between different applications and rapid network expansion are key aspects of Ethereum's development.

– Hyperledger: it is a platform that differs from Ethereum. It is a project-based, distributed ledger focused on blockchain applications that use smart contracts to build trust between all network members. Hyperledger aims to provide blockchain solutions for specific use cases in different domains, such as business, government and industry. Hyperledger's main objective is to develop modular technologies and frameworks that can be adapted and used in a variety of scenarios.

These two open-source platforms have different approaches, but they both play an essential role in the evolution and adoption of blockchain technology.

5.2.2. *Blockchain categories*

There are different types of blockchain, generally classified into three categories: public blockchain, private blockchain and consortium blockchain. Each type has its own characteristics and is geared to specific uses:

– Public blockchain: it is an open system accessible to anyone. The data available on this system are open and transparent. All members can participate in the registration and verification of information. In some cases, the consortium chain can be partially decentralized, allowing members to directly identify and select the database for reading authorizations. This type of network can be public and open to all or restricted to network members (Ismail and Materwala 2020a).

– Private blockchain: it is also known as a "permissioned blockchain"; it generally exists within an organization. Unlike public blockchains, private blockchains are closed networks that only allow the participation of certain authorized entities. In this case, participants generally have read and write rights, and private blockchains are administered by a network administrator who grants specific rights and restrictions to participants (Marwan et al. 2020).

– Consortium chain: it is a combination of the features of public and private blockchains. The consortium chain requires known participants, unlike the public blockchain, where participants are unknown. Unlike an open system, where anyone can validate blocks, or a closed system, where a single entity appoints block producers, a consortium blockchain involves many participants who function simultaneously as validators (O'Donoghue et al. 2019).

Each type of blockchain has its own specific advantages and use cases. Selecting the right type of blockchain depends on needs, data confidentiality and application requirements (Treiblmaier 2018; Helo and Shamsuzzoha 2020).

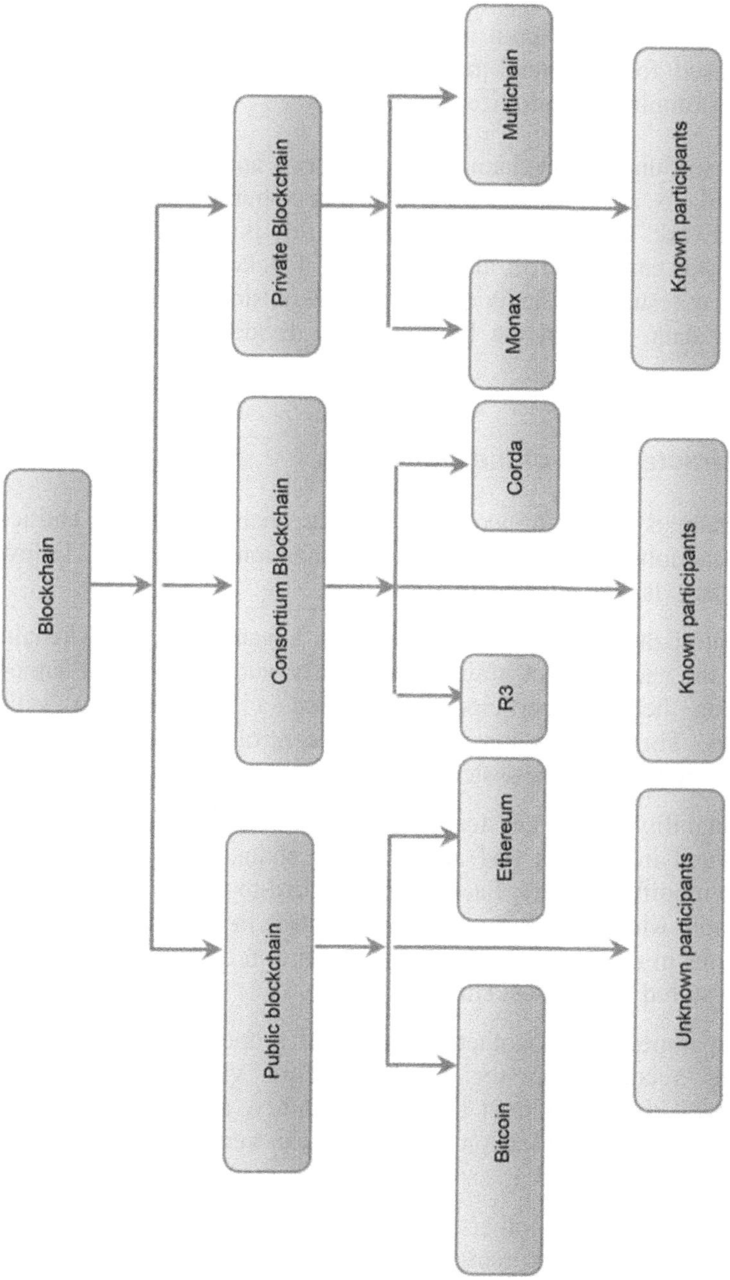

Figure 5.2. *Blockchain types and applications*

Figure 5.2 illustrates the different types of blockchain and their applications. The public blockchain is divided into two subsets, such as bitcoin and Ethereum. Bitcoin is defined as a peer-to-peer cryptocurrency primarily used for monetary transactions on the Internet. Ethereum is a platform for running decentralized applications.

The consortium's blockchain is based on two application types, R3 and Corda. Corda is an extension of R3, an open-source project designed to create a network. Corda is a platform whose role is to record and automate contracts. Data are stored in all parts of the blockchain, making it highly secure. As a result, any individual or entity outside the block can never modify the data. The private blockchain is divided into Multichain and Monax.

5.2.3. *Characteristics of the blockchain*

Blockchain is a peer-to-peer network that enables different entities to share transactions while ensuring decentralization. It has the following characteristics (Ismail and Materwala 2020b):

– Decentralization: it is one of the main advantages of blockchain. It eliminates the need for a centralized third party, such as a trusted authority, by enabling network participants to reach a consensus to validate transactions. This ensures that power and control are distributed among participants, reinforcing the system's resilience and security.

– Immutability: it is a key feature of blockchain. Once a transaction has been recorded in a block and added to the chain, it becomes extremely difficult to modify or delete. Each block is linked to the previous block by a hash function, which means that any modification in one block will result in modifications in all subsequent blocks. This reinforces the integrity and trust of the data stored on the blockchain.

– Transparency: it is another fundamental property of blockchain. All transactions recorded on the blockchain are visible to all network participants. This fosters trust and accountability, as each participant has access to a complete transaction history and can verify the accuracy of the information recorded.

– Traceability: it is another important feature of blockchain. As each transaction is recorded and replicated across the entire network, it becomes possible to follow the entire course of a transaction. This can be particularly useful in areas such as the supply chain, where it is important to know the origin, processing stages and transport conditions of products.

– Trust: blockchain fosters trust between participants who do not know each other. Thanks to consensus mechanisms such as proof-of-work or proof-of-stake, participants can agree on the state of the registry without having to rely on a central authority. This reinforces the trust and integrity of the network.

However, it is important to note that while blockchain offers many beneficial features, it is not the ideal solution for all use cases and can impose limitations and raise challenges. Considerations such as scalability, data confidentiality, implementation costs and energy consumption need to be considered when adopting blockchain.

5.3. Health sector

e-Health systems can be accessed by several actors: patients, doctors, insurers, pharmaceutical manufacturers and governments.

5.3.1. *Patients*

The patient is the main beneficiary of the healthcare industry. Healthcare policies must be designed with the patient's well-being in mind. Unfortunately, in recent years, there has been a shift from a patient-centered to a profit-driven approach within healthcare organizations. The interests of healthcare organizations, healthcare professionals and other healthcare providers are often represented by various government agencies or unions. However, there is no organization dedicated to defending patients' rights.

Thanks to blockchain technology, a system of trust and transparency is established between patients, healthcare professionals and medical organizations. Patients can share their data securely with healthcare professionals, promoting better care coordination. What is more, patients can control who has access to their data and protect their records from falsification.

Therefore, the adoption of blockchain technology in healthcare puts the patient back at the center of the system. This ensures easy and secure access to medical records, reinforcing the protection of patients' rights.

5.3.2. *Doctors*

Physicians play a crucial role in e-health systems. They are responsible for ensuring that patients receive the appropriate care they need.

Thanks to their knowledge and professional experience, doctors accurately assess patients' conditions. However, the integration of blockchain technology into the medical field has made their work considerably easier. Indeed, blockchain makes patient records available to them on a centralized platform, accompanied by summaries and reports. This increased accessibility and the ability to quickly interpret this information has greatly improved their efficiency.

Blockchain technology guarantees the integrity and security of medical data, enabling doctors to consult patient records with complete confidence. Thanks to this technology, they can make faster, more informed decisions, based on accurate, up-to-date information. This reduces the time needed to access relevant data and facilitates clinical decision-making.

The use of blockchain technology in the medical field has considerably simplified the work of doctors. It gives them rapid, secure access to patient records, accompanied by relevant summaries. This enables them to interpret information more effectively, resulting in more informed decisions and better-quality care for patients.

5.3.3. *Insurance sector*

Health insurance is a plan that covers all or part of a patient's medical expenses. The insurance company develops a financing structure based on the people covered, and this benefit is administered by organizations. Having insurance has become essential, as medical care has become expensive and chronic illnesses have become widespread throughout the world.

These days, insurance companies deal with a lot of paperwork and are prone to human error. There are also risks of scams and bankruptcies due to poor data management and falsification of information. Blockchain technology has made up for the lack of documentation. This technology has also accelerated the customer registration process, reduced costs and made data confidential and accessible only to authorized entities.

5.3.4. *Pharmaceutical industry*

Pharmaceutical companies play a vital role in the healthcare industry, since patients are dependent on the products that pharmaceutical companies manufacture. The industry discovers, develops, produces and sells medicines for patients. Companies are subject to various laws and regulations governed by private or governmental bodies. The prices of medicines are rising sharply and becoming more and more unaffordable. This is due to research, company investment in new drug discovery, marketing and supply chain expenses.

On the other hand, in recent times, the pharmaceutical industry has fallen victim to counterfeiting (Saxena et al. 2018). The consequences of this problem pose serious problems for patients' health. This phenomenon is very common in areas where monitoring and regulation are unreliable and need to be improved (Feng et al. 2019). Medicines pass through the supply chain and involve many intermediate suppliers before finally reaching the end user (the patient). A regulated body tests a batch of drugs, in order to know the quality of the medicines, but this system fails due to the difficulty of tracing the origin of each drug. Blockchain technology can ensure the traceability of pharmaceutical products from source to destination.

5.3.5. *Government*

The government's role in healthcare has expanded over time and is a major influence on power and political discourse in the healthcare industry (Straube 2013). The government has played a key role in promoting the use of preventive measures for chronic diseases and contributes to the efficiency of healthcare. It is the government's responsibility to protect and advance the interests of society by providing high-quality healthcare, and to serve the interests of patients by filling gaps where there are problems and inefficiencies (Tang et al. 2004). Blockchain has helped governments solve these problems.

5.4. Issues and challenges for the healthcare sector

Today's healthcare sector faces a constant challenge: improving the quality of care, reducing costs, increasing profits and securing data. Full digitization has brought major changes to healthcare services, enabling remote consultations via the Internet. Doctors can now remotely monitor patients' health using medical sensors based on the Internet of Things (IoT).

However, the collection of sensitive patient data by healthcare organizations makes them a major target for cybercrime. Centralized data storage systems are particularly vulnerable to cyber-attacks.

The introduction of blockchain technology offers a potential solution in the fight against cybercrime. It ensures the security and availability of data in real time, providing secure and up-to-date healthcare, guaranteeing personalized and authentic healthcare.

Decentralized distribution of healthcare information is one of the main advantages of blockchain (Koshechkin et al. 2021). Blockchain offers several services (Demirkan 2013; Ismail and Materwala 2020b).

5.4.1. *Quality*

Blockchain provides a transparent, secure and immutable framework for storing and sharing sensitive medical information. Thanks to its decentralized nature, blockchain eliminates traditional trusted third parties and enables healthcare professionals to access patient data quickly and securely. This promotes better care coordination and reduces the risk of errors due to missing or inaccurate information.

In addition, blockchain can reinforce the traceability of medicines and medical devices throughout their lifecycle, from manufacture to patient use. By recording every transaction on the blockchain, it becomes possible to accurately track the origin, provenance and authenticity of healthcare products. This reduces the risk of counterfeiting and distribution of out-of-date products and ensures greater safety for patients.

Finally, blockchain also facilitates the research and development of new treatments and therapies. Medical data are often dispersed in different systems and sharing them is limited by confidentiality concerns. Blockchain enables the creation of secure, interoperable data networks, where researchers and healthcare professionals can collaborate effectively to analyze datasets and obtain valuable information to improve care.

Therefore, the integration of blockchain into healthcare services offers numerous opportunities to improve the quality of care. By promoting transparency, security and collaboration, blockchain can help reduce errors, strengthen the traceability of healthcare products and stimulate medical research. Its adoption in the healthcare sector could therefore lead to significant improvements in the quality of care.

5.4.2. *Coordination*

The implementation of e-health systems encompasses a wide range of entities, such as patients, doctors, nurses, hospitals, pharmacies, external service providers, medical schools, insurers and manufacturers of pharmaceutical and biotech devices. In order to provide quality service, it is essential that all these entities cooperate and coordinate effectively. In this context, blockchain technology offers a promising solution to these problems.

Indeed, blockchain offers a secure and transparent environment for sharing sensitive healthcare information between different entities. Thanks to its decentralized nature, blockchain eliminates information silos and enables the entities involved to share data efficiently and securely. This fosters collaboration, communication and coordination between the various healthcare actors, helping to improve the quality of the services provided.

In this way, blockchain technology offers a promising solution for improving coordination and cooperation between entities in the healthcare sector. By promoting transparency, security and trust, blockchain can help deliver higher quality healthcare services by enabling more efficient information sharing, more accurate medication management and greater collaboration between the various actors.

5.4.3. *Integrity*

In the context of e-health systems, several entities use these systems and all need to ensure the integrity of the data they use. Blockchain technology can meet this need. When data is recorded on the blockchain, it is incorporated into a block that is linked to the previous block in the chain. This means that any modification made to a block will lead to a modification of all the blocks that follow it in the chain. This guarantees the immutability and integrity of recorded data. If data are mistakenly entered in a block, they are corrected by issuing a new transaction to record the correction.

The use of blockchain in e-health systems therefore makes it possible to guarantee the integrity of recorded data, by making it difficult to modify or falsify information once it has been recorded. This strengthens user confidence in the data available and promotes better quality of care by providing reliable and accurate information to healthcare professionals.

5.4.4. *Transparency*

One of the main advantages of blockchain technology is its transparency. When a change is made in the network, it is recorded as a transaction and becomes visible to all participants, who retain a copy of the record.

This transparency enables all network participants to verify and validate transactions independently. Each transaction is recorded in blocks that are added to the blockchain, creating an immutable history of all network activity.

This transparency makes it more difficult to manipulate or falsify data. Participants can consult the register to verify the accuracy of transactions and ensure that there are no unauthorized modifications.

As far as e-health systems are concerned, blockchain's transparency enables healthcare professionals, patients and other industry actors to verify the exactitude of medical information and track transaction history. This can help detect and prevent errors and fraud.

In this way, the transparency of the blockchain enables all participants to access a transparent and verifiable transaction history. This strengthens trust, facilitates data verification, and can help improve the system by detecting errors and preventing fraud.

5.4.5. *Traceability*

The traceability of operations carried out on data is a crucial aspect in e-health systems. Blockchain technology offers an effective solution for ensuring this traceability functionality.

When an operation is carried out on the blockchain, whether it involves the creation, modification or deletion of data, this operation is recorded in the form of a transaction.

This means that all operations carried out on data are recorded transparently and permanently on the blockchain. This makes it possible to trace the complete transaction history of each piece of recorded data, from its creation to its current state. Each user entity retains a copy of the register, enabling all participants to access and verify this information independently.

This traceability functionality offered by blockchain is extremely valuable in the healthcare sector. It makes it possible to track the evolution of medical data, detect unauthorized or malicious modifications and provide verifiable proof of all operations carried out on the data. This reinforces confidence in medical information and facilitates the implementation of regulatory compliance.

The traceability offered by blockchain is particularly beneficial for the management of drugs and medical devices. By recording all stages of the supply chain on blockchain, it becomes possible to trace origin, provenance, storage conditions, certifications and transactions. This helps guarantee the quality, safety and compliance of products used in healthcare.

In this way, blockchain technology ensures the traceability of operations carried out on data in e-health systems. The immutable blockchain makes it possible to track the complete history of transactions and provide verifiable proof of data integrity. This strengthens trust, facilitates the management of medication and medical devices, and eases regulatory compliance.

5.4.6. *Interoperability*

One of the major challenges in e-health systems is the management of multiple data sources from different entities and using a variety of formats. Blockchain technology can play a crucial role in facilitating communication and data exchange between these heterogeneous information systems.

Blockchain can be used as an interoperability layer to connect different systems and facilitate the secure transmission of data. Thanks to the decentralized nature of blockchain, entities can share their data transparently and securely, without depending on a central authority.

By using smart protocols and contracts, blockchain can normalize and standardize data formats, facilitating interoperability between different systems.

In addition, blockchain offers consensus features that enable participants to agree on the rules for managing and exchanging data. This establishes an environment of trust and guarantees the consistency of information shared between systems.

In this way, blockchain technology can help overcome data management challenges in e-health systems by offering the ability to communicate and exchange data between heterogeneous information systems. Blockchain can act as an interoperability layer, standardizing data formats and reinforcing trust in information exchange.

5.4.7. *Data sharing*

Secure data sharing is essential to providing effective treatment and care for patients. Collaboration and information exchange between different healthcare actors can help improve diagnostic accuracy and quality of care. Blockchain technology offers a solution for this secure data sharing.

In a blockchain network, data are stored in a decentralized, secure manner. Each transaction is signed (via a digital signature) and added to a block, which is then added to the blockchain. The data stored on the blockchain are protected by cryptographic mechanisms, making them resistant to falsification and unauthorized modification.

Thanks to this enhanced security, blockchain can facilitate data sharing between different healthcare actors, such as healthcare professionals, laboratories, insurers and patients. Access permissions can be managed on a granular basis, meaning that only authorized people can access specific data according to their needs and defined access rights.

5.4.8. *Costs*

Blockchain technology has the potential to help reduce costs in the healthcare sector, particularly with regard to the product development cycle and the tracking of medicines and medical devices.

The product development cycle in healthcare is long, regulated and costly. Research, clinical trials, regulation and marketing of medical products all require substantial investment. This can lead to high costs for pharmaceutical companies, medical device manufacturers and other actors in the medical industry.

What is more, blockchain can play a crucial role in tracking medicines and other healthcare devices. Thanks to the traceability offered by blockchain, it becomes possible to track the entire journey of a medicine or medical device, from its manufacture to its distribution and use by patients. This can help prevent counterfeiting, guarantee product authenticity, and detect quality problems more quickly, which can reduce the costs associated with product recalls and litigation.

In addition, the use of smart contracts can automate and streamline payment and reimbursement processes, reducing administrative costs and improving the efficiency of financial transactions in healthcare.

In this way, blockchain technology can help reduce healthcare costs by improving the transparency and efficiency of product development processes, enabling the tracking and tracing of specific medication and devices.

5.4.9. *Data volume*

The sheer volume of medical data in different formats represents a major challenge when it comes to sharing it between different entities. However, blockchain technology can play a crucial role in resolving these data volume issues.

Blockchain provides a distributed, decentralized infrastructure for efficiently storing and sharing large quantities of data. Medical data can be encapsulated in blockchain transactions and securely added to the blockchain. Each participant in the network has a copy of the registry, making it easy to access and share data.

In this way, blockchain technology offers potential solutions for managing the large volume of medical data in different formats. It enables secure, distributed data storage, facilitates interoperability and format conversion and provides control over patient consent. However, IT resourcing and scaling considerations must be taken into account to guarantee optimal performance.

5.5. Application of blockchain technology in e-health systems

Blockchain technology is attracting growing interest in healthcare, due to its ability to guarantee the security and confidentiality of medical data. Medical procedures involve the collection of data from a variety of sources, and blockchain offers a potential solution for authenticating and verifying the various parties involved (Ahmad et al. 2021).

The use of blockchain technology in healthcare systems has grown in popularity in recent years. A study conducted in the United States revealed a strong interest for these solutions among consumers of medical services. Some 19% of hospital executives and 80% of care recipients were considering or in the process of implementing blockchain solutions (Koshechkin et al. 2021).

The Covid-19 pandemic also accentuated the need to track and manage sensitive data securely. This has raised important issues of confidentiality, and various strategies have been explored to protect individual privacy. The adoption of blockchain technology has been seen as a potential solution due to its ability to guarantee data integrity and security while preserving confidentiality.

However, it should be noted that not all the conceptual ideas for using blockchain technology in healthcare have yet been implemented on a large scale. Research and development in this area is ongoing, and there is still much to be done to fully explore the potential of blockchain in medical applications.

Let us now take a look at the leading applications of blockchain.

5.5.1. *Electronic health records*

Electronic health records (EHRs) are one of the most popular areas where blockchain technology is being applied in healthcare. According to one study (Krishnan et al. 2020), around 41% of articles discussing blockchain technology in healthcare focus on EHRs.

e-Health records are essential components of digital healthcare systems, as they enable secure storage of patient medical data, appointment management, as well as billing and accounting management. As patient data are highly sensitive and confidential, it is crucial to implement robust security mechanisms to guarantee their integrity and confidentiality.

Blockchain technology offers functionalities that make it possible to secure EHRs. For example, by using blockchain, medial data can be stored in a decentralized and immutable way, reducing the risk of falsification or manipulation of records. What is more, the cryptography used in blockchain ensures data confidentiality by enabling authorizations control access.

By using blockchain for EHRs, it becomes possible to guarantee data authenticity, control access to patients' medical information and facilitate secure data sharing between different healthcare system stakeholders, such as doctors, pharmacies and insurers.

However, it should be noted that the widespread adoption of blockchain technology in EHRs requires additional efforts, not least in terms of regulation, standardization and interoperability between different healthcare systems. Nevertheless, blockchain applications in this field show great potential for improving the security and confidentiality of patient medical data.

5.5.2. *Pharmaceutical supply chain*

The pharmaceutical supply chain is a crucial area where the application of blockchain technology is seen as a major innovation. Drug counterfeiting is a serious problem that poses a major risk to patient health and safety, and blockchain technology can offer a promising solution to combat this scourge.

Drug counterfeiting is on the constant rise, particularly in developing countries, and is exacerbated by pandemics and global drug shortages. According to the World Health Organization, around 30% of medicines in developing countries are counterfeit.

Blockchain technology can help guarantee the authenticity of medicines and prevent counterfeit medicines from reaching patients. With traceability and quality control mechanisms integrated into the pharmaceutical supply chain, it is possible to track every stage of distribution, from production to delivery, by recording all transactions on the blockchain.

Blockchain enables the creation of a decentralized, immutable register of all medication transactions and movements, guaranteeing transparency and reliability of information. Manufacturers, distributors, pharmacies and other actors in the supply chain can access these data and verify the authenticity of medicines. This significantly reduces the risk of counterfeit medicines on the market and improves the quality of medical care.

Using blockchain, it also becomes possible to detect and report counterfeit medicines quickly, facilitating collaboration between health authorities, manufacturers and other stakeholders. Information stored on the blockchain, such as batch information, expiry dates and storage conditions, can be easily accessed and verified.

However, large-scale implementation of this technology requires collaboration between the various actors in the supply chain, as well as appropriate regulations and standards.

5.5.3. *Patient follow-up*

The combination of IoT technology and blockchain offers exciting opportunities for remote patient monitoring and healthcare data management. The IoT enables the collection of medial data from devices such as medical sensors, connected medical appliances, implants and so on. These data are then transferred to the blockchain network for secure storage, analysis and management.

By using IoT devices, patients can be monitored remotely, which can be particularly beneficial for patients with chronic illnesses or the elderly requiring continuous medical monitoring. The data collected, such as vital signs, glucose levels and blood pressure, is registered on the blockchain in a transparent and immutable way.

There are several advantages to using blockchain in this context. First, blockchain guarantees the security and confidentiality of healthcare data, as information is stored in a decentralized manner and protected by encryption mechanisms. Patients also have control over their own data and can grant selective access authorizations to healthcare professionals or other authorized parties.

What is more, blockchain enables greater traceability and transparency of medical data. Every transaction or addition of data is permanently recorded on the blockchain, creating a complete and verifiable history of information. This can be useful for medical research, error detection or data manipulation.

Finally, the use of blockchain facilitates the integration and interoperability of healthcare data from different sources. Patient data can be stored on a blockchain, enabling the various actors in the healthcare system, such as doctors, hospitals and insurers, to access relevant information securely and transparently. This can improve care coordination, medical decision-making and overall patient management.

However, it is important to consider the challenges of privacy, IoT device security and system interoperability to ensure successful and ethical implementation of this technology (Angraal et al. 2017; Engelhardt 2017).

5.5.4. *Scientific research in the health sector*

Blockchain technology can play an important role in medical research, particularly when it comes to clinical trials and protecting data confidentiality.

Clinical trials are a crucial step in the development of new medical treatments. However, they can often be prone to problems such as data falsification, manipulation of results or lack of transparency. Blockchain can help solve these problems by offering data traceability and immutability.

Using blockchain, clinical trial data can be stored transparently and securely, guaranteeing their integrity. Every step in the process, from data collection to results analysis, can be recorded on the blockchain, enabling independent verification and increased transparency. Researchers, regulators and clinical trial participants can thus have confidence in the accuracy and reliability of the data.

In addition, blockchain can help protect the confidentiality of individuals taking part in clinical trials. Sensitive personal data can be securely encrypted and stored on the blockchain, while enabling researchers to access the necessary information anonymously. This preserves the privacy of participants, while enabling data to be used for medical research.

In addition, blockchain can facilitate the management of consent in clinical trials. Participants can give their consent electronically using blockchain, guaranteeing the integrity and traceability of the consent process. Consents given can be permanently recorded on the blockchain, thus avoiding any subsequent disputes regarding authorizations that have been granted.

However, it is important to consider the challenges involved in adopting this technology, such as system interoperability, identity management and regulatory compliance.

5.5.5. *Analyzing medical data*

The application of blockchain technology in healthcare data analysis can be very promising, particularly when combined with artificial intelligence (AI).

AI can be used to extract valuable information from healthcare data, such as diagnoses, disease predictions or treatment recommendations. However, these data are often sensitive and need to be protected to ensure patient confidentiality (Mamoshina et al. 2018; Saxena et al. 2018).

By integrating blockchain technology, it is possible to create a secure environment for sharing and analyzing healthcare data. Blockchain enables data to be stored immutably and transparently, while ensuring confidentiality through encryption and access control mechanisms.

By combining AI with blockchain, it becomes possible to exploit these data in advanced ways. For example, AI algorithms can be used to analyze medical data and identify significant patterns, correlations or trends.

Using blockchain, the results of these analyses can be recorded in a transparent and traceable way. This not only guarantees the integrity of results, but also creates systems of validation and trust between the various actors in the healthcare field.

It should be noted, however, that the integration of these technologies also brings challenges, such as data quality and reliability, privacy protection and regulatory compliance.

These applications are just a few examples of the many other possibilities offered by blockchain technology in healthcare. Research and development in this field is constantly evolving, and new applications emerge regularly. Table 5.1 illustrates some of the applications of blockchain technology in the medical field (Carrano et al. 2022).

Applications	Benefits of using blockchain
Electronic health record	– Virtual health ID card – Easy retrieval of all health information
Supply of medicines	Efficiency and transparency of the chain
Patient follow-up	Secure processing of sensitive data retrieved from mobile devices
Research	Global and universal database for easy and immediate sharing
Analyzing medical data	Using artificial intelligence

Table 5.1. *Medical applications using blockchain*

5.6. Implementing blockchain technology in healthcare

Several projects have been developed to apply blockchain technology in the healthcare field. These projects aim to improve the scalability, decentralization and security of EHRs using distributed registries. Here are a few examples of projects commonly cited in the literature.

Projects using blockchain have used the Ethereum platform, such as MedRec, MedCredits, MIStore, RoboMED and other projects have used the Hyperledger platform, such as Healthchain and Medical Chain.

5.6.1. *MedRec*

The MedRec project, proposed by Azaria, Ekblaw and their collaborators, is one of the functional prototypes to apply blockchain technology to EHRs. They used the Ethereum platform to develop their system, called MedRec. MedRec's main objective is to control access to medical data and authenticate users. The system manages health records in a distributed way, enabling greater system interoperability and improving patients' ability to access their data. In addition, MedRec aims to improve data quality and quantity by encouraging patients to share their medical information securely.

The choice of the Ethereum platform enables the use of smart contracts to automate certain system functionalities, thus guaranteeing a high level of security and transparency in the management of health records (Azaria et al. 2016).

5.6.2. *MedCredits*

The MedCredits project is a system based on the Ethereum platform that aims to help doctors in the field of dermatology diagnose patients remotely, using e-health services. MedCredits' main objective is to provide a secure and reliable system using reputation-based mechanisms.

The MedCredits system implements reputation systems to encourage honest behavior and deter dishonest behavior. These reputation mechanisms help maintain trust between doctors and patients by assessing the quality of services provided. Doctors who provide accurate and reliable diagnoses are rewarded, while those who provide erroneous or misleading information may be penalized.

MedCredits also uses a service called Token-Curated Registry (TCR) to validate doctors' licenses and ensure that only legitimate doctors can join the platform. This blockchain-based validation process guarantees the integrity and competence of the healthcare professionals on the platform.

Ethereum-based smart contracts are used to automate payments between doctors and patients. Doctors can access patients' health data, diagnose their condition and prescribe treatments, all using blockchain to secure and share medical information confidentially (Mannaro et al. 2018).

5.6.3. *MIStore*

The MIStore project is an application of blockchain technology in the field of health insurance. It aims to improve the efficiency of the medical information storage process and facilitate exchanges between patients, medial institutions and insurers.

MIStore is using the Ethereum platform to deploy its blockchain-based storage system. Patients' medical information, such as health records, invoices and insurance claims, is stored securely and transparently on the blockchain. This enables the various entities involved in the health insurance process to quickly access the information they need to settle claims and authorize payments to patients.

Thanks to blockchain technology, MIStore offers several advantages. First, it eliminates the inefficiencies associated with paper-based processes and slow information exchanges. Medical data are recorded in an immutable and verifiable way, reducing errors, fraud and potential litigation.

What is more, blockchain's transparency enables insurers to quickly verify the accuracy of claims and speed up the settlement process. Payments to patients can be authorized more quickly, improving patient experience and insurer satisfaction (Zhou et al. 2018).

5.6.4. *Robomed*

Robomed is a network of clinical organizations that uses Ethereum blockchain-based smart contracts to deliver efficient medical services to patients. Healthcare organizations can register, connect and manage themselves within the Robomed network using Ethereum contracts.

The Robomed EHR module enables real-time monitoring of patient interactions, medical decision-making, staff access rights management and e-health consulting services. Patients can use the Robomed mobile module for remote consultations, appointment management and secure health data sharing with clinics. The RBM Ethereum token is used as a payment method in all Robomed organizations (Feng et al. 2019).

5.6.5. *HealthChain*

HealthChain is a Hyperledger-based data management platform for cancer radiotherapy sessions. The platform enables patients and providers to access medical records securely. HealthChain focuses on the trust and security of transactions between patients, physicians and specialists, offering a medical service based on trusted relationships (Hylock et al. 2019; Dubovitskaya et al. 2020).

5.6.6. *Medicalchain*

Medicalchain uses the Ethereum and Hyperledger Fabric platforms to provide secure remote consultation services. Patients can share their health data securely with healthcare professionals, in accordance with specified conditions. Medicalchain uses Hyperledger Fabric functionalities to implement access control policies. The ERC20 token is used on the

Ethereum platform to facilitate payments and identify insurance fraud (Albeyatti 2018).

Project	Year	Platform	Objectives	Targeted security service
MedRec	2016	Ethereum	Managing records in a distributed way	Authentication and confidentiality
MedicalChain	2018	Hyperledger	Secure storage of medical records	Authentication, integrity and confidentiality
HealthChain	2019	Hyperledger	Data management for cancer radiotherapy sessions	Authentication and confidentiality
Robomed	2017	Ethereum	Real-time monitoring of patient interactions	Authentication and confidentiality
MIStore	2018	Ethereum	Efficient data storage in the medical insurance sector	Authentication and confidentiality
MedCredits	2018	Ethereum	Data management in dermatology	Authentication and confidentiality

Table 5.2. *Characteristics of blockchain projects in the medical field*

These projects are all examples of the application of blockchain technology in healthcare. They exploit blockchain's functionalities to guarantee security, transparency, interoperability and efficient management of medical data, which can improve healthcare services and trust between the various actors.

Table 5.2 summarizes the main features of these projects.

5.7. Contribution of the blockchain solution

The blockchain provides a secure, permanent record of transactions between users without the intervention of a third party. The blockchain does not store data in a centralized database controlled by one party, but all blockchain entities have a synchronized copy of the data and can access it in the same way. Data integrity is ensured by hashing algorithms.

The decentralization feature increases the overall robustness of e-health systems, and thus patient EHRs are preserved against attack and accidental data loss (Demirkan 2013; Albeyatti 2018; Tasatanattakool and Techapanupreeda 2018; Carrano et al. 2022). The immutability of healthcare records is ensured by public-key cryptography, which means that each transaction must first be signed before it can be verified. Healthcare requirements, such as rapid, real-time sharing of EHRs (Mamoshina et al. 2018), patient-centric health data management (Saxena et al. 2018), cost reduction and service availability (Feng et al. 2019), can be met thanks to blockchain technology.

By harnessing the benefits of blockchain, many projects and initiatives seek to improve the efficiency, security and transparency of healthcare systems. However, it is important to note that despite its potential, blockchain technology is not a silver bullet and also presents challenges and limitations that need to be taken into account when implementing it. Table 5.3 presents a comparison of approaches to data processing in the medical field. The traditional approach refers to the non-digitized approach to care. The client/server approach refers to centralized architectures.

Blockchain technology offers several potential advantages over healthcare systems based on client/server architecture (Justinia 2019; Ismail and Materwala 2020b).

First, fault tolerance is improved thanks to blockchain. In client/server systems, where data are stored in a centralized database, a breakdown or loss of data can have serious consequences. On the other hand, blockchain is fault-tolerant thanks to its decentralized nature and the replication of data throughout the network. So, even if a node or part of the network fails, the data remain accessible and intact.

Data sharing is also simplified thanks to blockchain. In client/server systems, patient data are often scattered across the databases of different hospitals, making information sharing complex. Using a blockchain-based platform, patient data are stored and replicated across the entire network, facilitating secure data sharing between healthcare system entities.

System interoperability is another advantage of blockchain. In client/server systems, different entities often use different data formats and structures, leading to interoperability problems. By using a record synchronized and replicated in the blockchain, these problems can be solved, as all entities in the network have access to a common, shared data format.

	Traditional application	Client/server application	Application based on blockchain technology
Cost	High	Low	Low
Wait time for patients	High	Low	Low
Requirement for an in-person visit	Yes	No	No
Access	–	Only authorized users have access	Anyone can validate transactions on the network
System administration	Centralized	Centralized	Decentralized
Confidentiality and data security	Difficult to ensure	Difficult to ensure	By definition
Integrity	Difficult to ensure	Difficult to ensure	By definition
Transparency	No	No	Yes
Consensus	–	Validation in two phases	The majority of pairs agree on the outcome of transactions
Replication	–	Only the central part has a copy	Complete block replication on each pair
Identity theft	–	Accounts are often hacked	Encryption and anonymity, which makes it difficult to hack the blockchain

Table 5.3. *Comparison of data processing approaches in the medical field*

The management of medical examinations is simplified thanks to blockchain. In traditional systems, patient data are often dispersed between several healthcare providers, which can lead to additional medical costs and repeated medical examinations. Using blockchain, medical information is accessible to all authorized actors, avoiding duplication of tests and examinations, thereby reducing costs and adverse effects on the human body.

Finally, data security is enhanced by the application of blockchain technology in e-health systems. Blockchain uses advanced cryptographic mechanisms to secure transactions and data, guaranteeing the secure exchange of information between healthcare system entities. What is more, the decentralized nature of blockchain makes attacks and data falsification more difficult.

Thus, blockchain technology offers several potential advantages over client/server-based healthcare systems, including improved fault tolerance, simplified data sharing, enhanced interoperability, more efficient management of medical examinations and enhanced data security. However, it is important to take into account the challenges and considerations involved in implementing blockchain in healthcare.

5.8. Conclusion

In conclusion, blockchain technology has enormous potential to transform the healthcare sector by offering innovative and secure solutions. It offers benefits such as data transparency, security, decentralization and improved efficiency of healthcare services.

However, it is important to recognize that the adoption of blockchain in healthcare is still at an early stage and presents challenges such as regulation, interoperability of existing systems and patient privacy. Collaboration between healthcare actors, governments and technology experts will be essential to overcome these challenges and fully exploit the potential of blockchain technology.

Despite these challenges, blockchain applications in healthcare are promising and create new opportunities to improve healthcare, facilitate medical research, strengthen trust between stakeholders and deliver more efficient services to patients.

It is crucial to continue exploring and developing the uses of blockchain in healthcare, while remaining mindful of privacy, security and ethical issues. With a collaborative approach and concerted efforts, blockchain technology can contribute to a positive transformation of the healthcare sector, offering innovative solutions and improving the quality of medical services.

5.9. References

Ahmad, R.W., Salah, K., Jayaraman, R., Yaqoob, I., Ellahham, S., Omar, M. (2021). The role of blockchain technology in telehealth and telemedicine. *International Journal of Medical Informatics*, 148, 104399.

Albeyatti, A. (2018). Whitepaper. Medicalchain.

Angraal, S., Krumholz, H.M., Schulz, W.L. (2017). Blockchain technology: Applications in health care. *Circulation: Cardiovascular Quality and Outcomes*, 10(9), e003800.

Azaria, A., Ekblaw, A., Vieira, T., Lippman, A. (2016). Medrec: Using blockchain for medical data access and permission management. In *2016 2nd International Conference on Open and Big Data (OBD)*. IEEE.

Carrano, F.M., Sileri, P., Batt, S., Di Lorenzo, N. (2022). Blockchain in surgery: Are we ready for the digital revolution? *Updates in Surgery*, 74, 1–4.

Dellys, S. and Benbouabdellah, S. (2020). Applications de la technologie blockchain. PhD Thesis, Université Akli Mohand Oulhadje-Bouira, Bouira.

Demirkan, H. (2013). A smart healthcare systems framework. *It Professional*, 15(5), 38–45.

Dubovitskaya, A., Baig, F., Xu, Z., Shukla, R., Zambani, P.S., Swaminathan, A., Jahangir, M.M., Chowdhry, K., Lachhani, R., Idnani, N. et al. (2020). ACTION-EHR: Patient-centric blockchain-based electronic health record data management for cancer care. *Journal of Medical Internet Research*, 22(8), e13598.

Engelhardt, M.A. (2017). Hitching healthcare to the chain: An introduction to blockchain technology in the healthcare sector. *Technology Innovation Management Review*, 7(10).

Feng, C., Li, N., Wong, M.H., Zhang, M. (2019). Initial coin offerings, blockchain technology, and white paper disclosures. *SSRN Electronic Journal*. doi: 10.2139/ssrn.3256289.

Helo, P. and Shamsuzzoha, A.H.M. (2020). Real-time supply chain – A blockchain architecture for project deliveries. *Robotics and Computer-Integrated Manufacturing*, 63, 101909.

Hylock, R.H. and Zeng, X. (2019). A blockchain framework for patient-centered health records and exchange (HealthChain): Evaluation and proof-of-concept study. *Journal of Medical Internet Research*, 21(8), e13592.

Ismail, L. and Materwala, H. (2020a). Blockchain paradigm for healthcare: Performance evaluation. *Symmetry*, 12(8), 1200.

Ismail, L. and Materwala, H. (2020b). BlockHR: A blockchain-based framework for health records management. In *Proceedings of the 12th International Conference on Computer Modeling and Simulation*.

Justinia, T. (2019). Blockchain technologies: Opportunities for solving real-world problems in healthcare and biomedical sciences. *Acta Informatica Medica*, 27(4), 284.

Koshechkin, K., Lebedev, G., Radzievsky, G., Seepold, R., Martinez, N.M. (2021). Blockchain technology projects to provide telemedical services: Systematic review. *Journal of Medical Internet Research*, 23(8), e17475.

Krishnan, S., Balas, V.E.E., Golden, J., Robinson, Y.H., Balaji, S., Kumar, R. (eds). (2020). *Handbook of Research on Blockchain Technology*. Academic Press, Cambridge.

Mamoshina, P., Ojomoko, L., Yanovich, Y., Ostrovski, A., Botezatu, A., Prikhodko, P., Izumchenko, E., Aliper, A., Romantsov, K., Zhebrak, A., Ogu, I.O., Zhavoronkov, A. (2018). Converging blockchain and next-generation artificial intelligence technologies to decentralize and accelerate biomedical research and healthcare. *Oncotarget*, 9(5), 5665.

Mannaro, K., Baralla, G., Pinna, A., Ibba, S. (2018). A blockchain approach applied to a teledermatology platform in the Sardinian region (Italy). *Information*, 9(2), 44.

Marwan, M., Temghart, A.A., Sifou, F., AlShahwan, F. (2020). A decentralized block-chain-based architecture for a secure cloud-enabled IoT. *Journal of Mobile Multimedia*, 389–412.

Nakamoto, S. and Bitcoin, A. (2008). A peer-to-peer electronic cash system. Bitcoin [Online]. Available at: https://bitcoin.org/bitcoin.pdf.

O'Donoghue, O., Vazirani, A.A., Brindley, D., Meinert, E. (2019). Design choices and trade-offs in health care blockchain implementations: Systematic review. *Journal of Medical Internet Research*, 21(5), e12426.

Saxena, M. and Sanchez, R. (2018). Method for providing healthcare related. Report, Blockchain-Associated Cognitive Insights Using Blockchains.

Straube, B.M. (2013). A role for government: an observation on federal healthcare efforts in prevention. *American Journal of Preventive Medicine*, 44(1), S39–S42.

Tang, N., Eisenberg, J.M., Meyer, G.S. (2004). The roles of government in improving health care quality and safety. *The Joint Commission Journal on Quality and Safety*, 30(1), 47–55.

Tasatanattakool, P. and Techapanupreeda, C. (2018). Blockchain: Challenges and applications. In *2018 International Conference on Information Networking (ICOIN)*. IEEE.

Theodouli, A., Arakliotis, S., Moschou, K., Votis, K., Tzovaras, D. (2018). On the design of a blockchain-based system to facilitate healthcare data sharing. In *17th IEEE International Conference on Trust, Security and Privacy in Computing and Communications/12th IEEE International Conference on Big Data Science and Engineering (TrustCom/BigDataSE)*.

Treiblmaier, H. (2018). The impact of the blockchain on the supply chain: A theory-based research framework and a call for action. *Supply Chain Management: An International Journal*, 23(6), 545–559.

Xie, Y., Zhang, J., Wang, H., Liu, P., Liu, S., Huo, T., Duan, Y.-Y., Dong, Z., Lu, L., Ye, Z. (2021). Applications of blockchain in the medical field: Narrative review. *Journal of Medical Internet Research*, 23(10), e28613.

Zhou, L., Wang, L., Sun, Y. (2018). MIStore: A blockchain-based medical insurance storage system. *Journal of Medical Systems*, 42(8), 149.

6

Using Biometrics to Secure Intra-BAN Communications

Omessaad HAMDI[1], **Mohamed AYMEN CHALOUF**[2] **and Amal SAMMOUD**[3]

[1] *IEEE, Rennes, France*

[2]*IRISA, Université de Rennes, France*

[3] *MIRA, Aflokkat, Ajaccio, France*

6.1. Introduction

The evolution of information and communication technologies (ICT) has led to the emergence of e-health systems. These systems are generally based on wireless body area networks (WBANs) to provide remote patient monitoring. The data processed and exchanged by the WBAN's medical sensors are sensitive and confidential. For this reason, it is vital to secure the transmission of these data within the WBAN.

In this chapter, we focus on the use of biometrics to secure communications between the WBAN's medical sensors. We will study the most interesting security approaches, based on the use of electrocardiogram (ECG) as biometric data, to secure exchanges within the WBAN.

This chapter is structured as follows. Section 6.2 presents the security of WBAN. In this section, we present the architecture of this type of network, as well as the security requirements to resist various attacks. In section 6.3, we introduce the different security solutions for communications within the WBAN (intra-WBAN or intra-BAN). Section 6.4 presents biometric-based security solutions for intra-WBAN communications. In section 6.5, we

e-Health Security Management,
coordinated by Omessaad HAMDI.

present an interpretation and comparison of the security approaches studied. Section 6.6 concludes the chapter.

6.2. Security for WBAN

6.2.1. *Architecture of an e-health system*

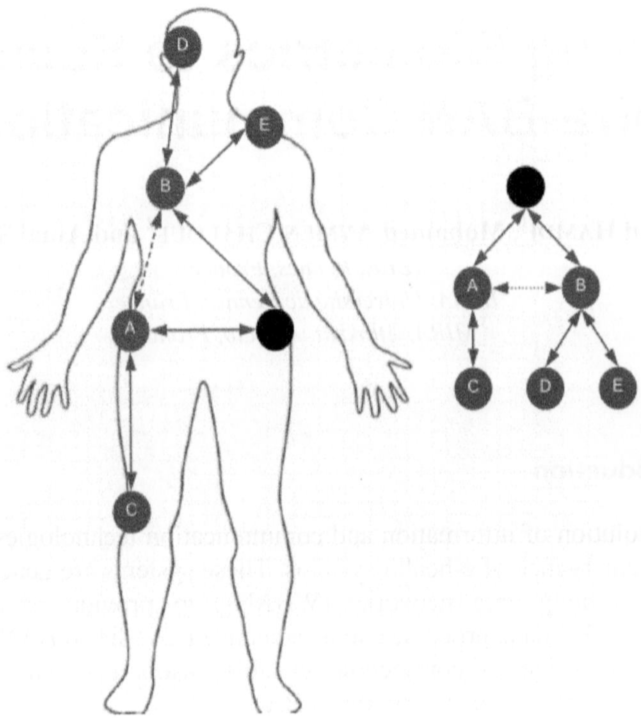

Figure 6.1. *Example of a BAN (Hajar et al. 2021)*

To understand the security mechanisms used in a WBAN, we first need to understand the structure and architecture of the e-health systems in which these networks are used. Indeed, e-health systems are generally made up of a set of wireless medical sensors fitted to the patient, a gateway (smartphone, medical box, etc.), remote servers (storage and processing) and healthcare professionals, with the necessary rights to access patient medical data. Generally speaking, these systems involve three main types of communication (Figure 1.3):

– Level 1: intra-BAN communications. This level includes communication between the sensors on the WBAN to measure and process the patient's medical signals (Figure 6.1), as well as the transfer of data collected by these sensors to the personal server (e.g. PDA).

– Level 2: inter-BAN communications. This level includes communications between the personal server and the access points. In effect, the personal server acts as a gateway for real-time transmission, via access points, of data collected by WBAN to a remote medical server.

– Level 3: beyond-BAN communications. This level covers communications between access points and medical servers via the Internet.

In this chapter, we focus on intra-BAN, or more specifically, intra-WBAN communications.

6.2.2. *Security requirements for WBANs*

WBANs require certain security measures to guarantee, among other things, the protection of patient privacy. Indeed, as WBANs are generally used to transmit sensitive and private information on a patient's state of health, it is essential to protect this information from unauthorized accessor falsification, which could endanger the patient's life. In this context, we present below the main security requirements for the proper acceptance of these new technologies (Narwal et al. 2021):

– Data confidentiality: data confidentiality refers to the protection of data against unauthorized disclosure.

– Data integrity: data integrity refers to measures taken to protect the content of a message against external modifications.

– Data freshness: data freshness techniques ensure that old data are not retransmitted.

– Data authentication: WBAN nodes must be able to verify that information communicated in the network is shared by a trusted node (another node belonging to their network).

– Reliability: WBANs must ensure the reliability and security of shared information. Indeed, any failure to retrieve data can put the patient's life at risk.

6.2.3. *WBAN attacks*

WBANs are exposed to several active and passive attacks. The most common attacks (Kompara et al. 2018) are as follows:

– Passive eavesdropping: the aim of this attack is to gain access to data exchanged in the network in order to discover important information. The sensitive data collected can be used by the attacker in an illegal way.

– Identity theft: generally speaking, the theft of a WBAN node's identity enables an attacker to present themself as a node belonging to the network. To do this, the attacker obtains private credentials and uses them to impersonate a legitimate network node. An attacker who succeeds in this attack will obtain the same rights as the network nodes.

– Replay: this attack consists of capturing a sent message and sending it back at a later time. In this way, the attacker can cause the receiver to run out of energy after sending a certain number of messages.

– Man-in-the-middle (MITM): the attacker can intercept and relay messages between two network nodes in direct communication. This enables the attacker to listen in and manipulate the information transmitted in real time.

– Compromised node: this attack enables an attacker to extract all data from a compromised node (including secret keys) and use it to gain unauthorized access to the network.

– Session key disclosure: this attack consists of obtaining a valid session key, used by two nodes on the same WBAN, to secure their communications. This enables the attacker to access the content of these communications.

6.3. Security solutions for intra-WBAN communications

The security of intra-WBAN communications is very important, since sensitive and private medical information is involved. Many security solutions have been proposed to protect these communications.

We present below the most used solutions for intra-WBAN security.

6.3.1. *TinySec*

TinySec is a solution for authenticating and encrypting data in WBANs (Al-janabi et al. 2017). In this solution, a shared group key is used by all nodes in the WBAN to encrypt communicated data. This group key is manually programmable before the network is deployed (Al-janabi et al. 2017).

6.3.2. *Biometric methods*

Biometric methods are used to secure intra-WBAN communications (Al-janabi et al. 2017). Biometric-based security solutions involve using physiological signals measured by sensor nodes in the WBAN as a means of managing the encryption keys used in the network. Typically, these physiological signals are unique to each individual, possessing a random character and measurable by all the sensors in the WBAN (Narwal et al. 2021).

6.3.3. *ZigBee security*

ZigBee was created to manage short-range wireless communications for very low-power devices (Narwal et al. 2021). The ZigBee network layer defines security services (Al-janabi et al. 2017). In particular, ZigBee provides mechanisms guaranteeing secure communications, establishment protection, as well as cryptographic key distribution, frame encryption and device control.

6.3.4. *Bluetooth security*

It includes various protocols, such as baseband, Link Management Protocol (LMP) and Logical Link Control and Adaptation Protocol (L2CAP). Baseband enables Bluetooth devices to link up and exchange data in packet form. The LMP protocol is responsible for security, such as data encryption, authentication and encryption key exchange. L2CAP can support a higher level of multiplexing and packet reassembly, which can help provide communication with a certain quality of service (Al-janabi et al. 2017).

6.3.5. *Elliptical curve cryptography*

Elliptic curve cryptography (ECC) offers the option of realizing public key cryptography in WBANs (Narwal et al. 2021). Indeed, ECCs are well suited to secure data communication and key dissemination in resource-limited environments (Narwal et al. 2021).

Table 6.1 compares the security solutions for intra-WBAN communications presented in this section.

Security solutions	Performances	Limitations
Bluetooth security	– Confidentiality and authentication	– It does not support authorization, integrity control or non-repudiation. As a result, the Bluetooth security solution is unsuitable. – Significant energy consumption, due to the radio's longer activation period.
Biometric methods	– Reasonable resource consumption Keys are sufficiently random	– Costs and calculations depend on the capacity of the sensor used (implantable/portable).
Elliptical curve cryptography (ECC)	– Increased security	– Higher energy consumption than symmetrical cryptography.
TinySec security	– Confidentiality and authentication	– Achieves minimal security. Not secure against attack by compromised node.
ZigBee security	– Authentication and key exchange – Energy efficiency	– Low data transmission speed and network stability.

Table 6.1. *Security solutions for intra-WBAN communications (Narwal et al. 2021)*

The sensors of the WBAN often have very limited computing, memory capacities and energy. These constraints make establishing adequate security a major challenge. Indeed, Intra-WBAN communications must be secured while respecting the resource constraints of medical sensors. As shown in Table 6.1, the solutions presented have drawbacks in terms of security and/or the limited capabilities of wireless medical sensors.

In the following, we will focus on security solutions based on biometric data. As the WBAN is made up of sensors implanted around or inside the human body, the idea of taking advantage of physiological signals to secure this network has emerged. Biometric data can be used to generate symmetrical cryptographic keys that are sufficiently random to withstand multiple attacks on WBANs. In the remainder of this chapter, we present security protocols based on biometric data, which aim to provide a good level of security (robustness against the best-known attacks) while adapting to the constraints of WBANs (energy, memory and computation).

6.4. Biometric data-based security solutions for WBANs

6.4.1. *Biometrics*

The WBAN consists of a set of interconnected medical sensors. These sensors are often very limited in terms of computing capacity, memory and energy. Therefore, the most important challenge for safety protocols is to minimize resource consumption in order to maximize the lifetime of these sensors.

Biometric methods (solutions based on biometric data, such as ECG) have been shown to be well suited for securing intra-WBAN communication, as they minimize resource consumption (Pirbhulal et al. 2019). This is made possible by the following characteristics of biometric data (Morales et al. 2019):

– universality: present for every person;

– uniqueness: different from one person to another;

– collectability: easy to collect and measure;

– availability: suitable for devices with low computing power.

These physiological values can be used to provide authentication for WBAN sensors (Narwal et al. 2021). In addition, biometric data are well suited for symmetric cryptography key generation and/or sharing (Ali et al. 2015; Al-Janabi et al. 2017; Masdari et al. 2017; Kompara et al. 2018; Arya et al. 2020; Hussain et al. 2021; Narwal et al. 2021). These biometric methods include the ECG signal and the photoplethysmogram (PPG) signal.

The availability of ECG or PPG signals, similar for all nodes of a single WBAN, is a key constraint for the efficient establishment (generation and sharing) of symmetric cryptography, as well as for authentication. Indeed, the two most widely used methods for ensuring this availability are time synchronization and network broadcasting. For the first method, all nodes of the WBAN must periodically and simultaneously measure a sequence of ECG/PPGs. In the second method, one sensor in the WBAN will be responsible for periodically measuring the ECG/PPG sequence and broadcasting it to the other nodes in the WBAN (Sammoud et al. 2020).

6.4.1.1. *Electrocardiogram*

An ECG (or EKG) signal is a representation of the heart's electrical activity. The device used to measure the ECG is called an electrocardiograph. This device records the cardiac electrical current to obtain a time–voltage graph of the cardiac rhythm.

There are two approaches to extract biometric features from the ECG signal: fiduciary and non-fiduciary. The ECG signal is made up of three main waves: P, QRS and T (Figure 6.2). The fiduciary approach detects, processes and classifies the peaks and intervals between these three waves. The ECG is generally characterized by the randomness of the IPI (inter-pulse interval). This interval represents the time separating two successive R peaks. In contrast, the non-dual approach processes the ECG signal based on the waveform in the frequency domain.

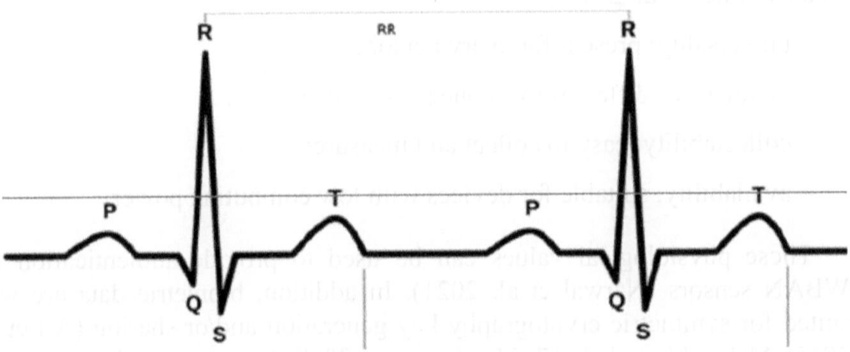

Figure 6.2. *An ECG sequence (Yao et al. 2011)*

6.4.1.2. *Photoplethysmogram*

A PPG signal is a signal obtained from the observation of variations in blood volume in a tissue of a living being. This signal is characterized by its variance over time. Like the ECG signal, temporal information extracted from the IPI interval can also be derived from the PPG signal (Poon et al. 2006). Figure 6.3 illustrates a PPG signal sequence.

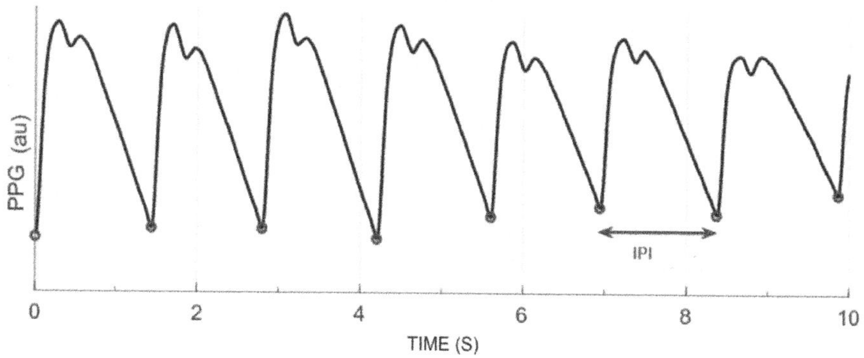

Figure 6.3. *A PPG sequence (Charlton et al. 2020)*

6.4.2. *Examples of security approaches for intra-WBAN communications using biometrics*

A number of research projects have focused on the use of biometrics to secure intra-WBAN communications.

ECG and PPG have generally been used to generate and/or share the symmetrical keys used to encrypt WBAN exchanges, thanks to their random nature. Indeed, Pirbhulal et al. (2019) have demonstrated that the ECG signal enables the generation of reliable and robust symmetric keys with reasonable resource consumption.

In 2007, Bui et al. (2007) proposed a biometric-based key distribution protocol named as single-point fuzzy key management with ECG. This protocol involves a specific WBAN node generating a symmetrical key using a cryptographic key generator. This node then distributes the key to other WBAN nodes using fuzzy commitment and hash functions.

Bui et al. (2007) have also proposed a second approach to ECG-based session key generation, called multipoint fuzzy key management. This approach uses an ECG signal, an error correcting code (ECC) and a morphing function to generate and distribute a session key.

Mana et al. (2009) presented an approach to session key establishment (generation and distribution) in WBANs, called SEKEBAN (secure and efficient key exchange for wireless body area network). This approach consists of establishing, at the outset, session keys to secure exchanges between WBAN nodes and a central node. This secures the communications needed to generate the ECG-based session keys that will later be used to secure intra-WBAN communications.

Yao et al. (2011) proposed an approach for establishing ECG-based session keys to secure intra-WBAN communications. The generated keys are used to protect the data communication exchanged between each sensor in the WBAN and the central node.

Zaghouani et al. (2015) proposed the ELPA (ECG linear prediction key agreement scheme) approach. This approach consists of using the LPC (linear prediction coding) method in order to protect the sharing of the biometric session key, generated on the basis of the ECG signal, between the nodes involved.

They also proposed a biometric-based security protocol named physiological feature-based key agreement (PFKA). This protocol aims to secure communications between nodes in a WBAN by enabling two nodes to generate an ECG-based session key and share it securely, using hash functions and ECC.

Cho et al. (2016) proposed an approach for establishing PPG-based generated session keys and bloom filter to secure WBANs. They claimed that the bloom filter technique minimized the computational cost of their approach.

They also proposed another approach to securing intra-WBAN communications, whose main advantage is to minimize the time required to generate session keys. This approach is based on the PPG signal and uses ECC.

Sammoud et al. (2020) have proposed a biometric (ECG)-based session key establishment protocol to secure intra-WBAN communications. The proposed protocol involves a third WBAN node and uses an ECC and a morphing function. Ali et al. (2015) have shown that their proposal minimizes the resource consumption of the WBAN nodes while guaranteeing the security of the key exchange mechanism.

6.4.3. *The approach of Sammoud et al.*

The security protocol proposed in Sammoud et al. (2020) involves using the ECG as biometric data in order to generate and distribute symmetrical keys. These keys will subsequently be used to secure intra-WBAN communications. Reliable distribution of a session key between two WBAN nodes was ensured via a third node. The Sammoud et al. protocol considers a mesh-tree topology and makes use of a third node, which is none other than the closest common ascendant of the two nodes wishing to communicate. Initially, it is assumed that the nearest ascendant (third node) is the parent node of both ends of a communication.

Notations	Explications
N_i	A WBAN intermediate node
N_{ip}	The father node of the N_i node
ECG_i	The ECG signal measured by N_i
U_i	The binary code for ECG_i
k_i	The session key chosen during the first phase (establishment of a symmetrical key between father and child (128 bits))
k_{pi}	The primary key generated by N_i (128 bits)
$k_{sessionij}$	The symmetric key generated from biometric data by N_i and N_j (128 bits)
E_k (.)	Symmetric key-based encryption function k
D_k (.)	Decryption function based on the symmetric key k
M (.)	Morphing function
a || b	Concatenation between two binary sequences a and b
$a \oplus b$	XOR function of two binary sequences a and b

Table 6.2. *Notations used in the description of Sammoud et al.'s (2020) approach*

This protocol consists of two phases:

– Father-child symmetric key establishment phase: with the exception of the root node (usually the central node), each WBAN node must share a symmetric key with its father. This key will be used to secure father child communications, such as the data used to establish the key generated from biometric data.

– Symmetric key generation phase using biometric data between two nodes: the generated key will be used to secure intra-WBAN communications.

The notations used in the description of the approach presented in Sammoud et al. (2020) are summarized in Table 6.2.

6.4.3.1. *Establishing a symmetrical key between father and child*

In this phase, the child node chooses, arbitrarily, a symmetric key k_i from the pre-stored set of cryptographic keys. In reality, this set of keys is generated by all nodes, during the initialization phase of the intra-WBAN, using standard key generation functions (Cho and Chung 2017). The key k_i is then transmitted to the parent node using a fuzzy commitment function (An 2012). For this phase to function correctly, the father node and its child node must have an ECG signal measured at the same time. At the child node, the ECG and its temporary identifier *Idt* will be used, in addition to the commit function F(.,.), to securely communicate the symmetric key. This is done as follows:

$$A_i = h(U_i \,||\, Idt_i) \qquad [6.1]$$

$$F(k_i, A_i) = (M(k_i) \,||\, A_i \oplus k_i) \qquad [6.2]$$

On the father node's side, the symmetrical key of the received message is recovered by using an ECG signal, the temporary identifier of the child node and an ECC f(.) to decode the key:

$$A'_i = h(U_{ip} \,||\, Idt_i) \qquad [6.3]$$

$$k'_i = f(A'_i \oplus A_i \oplus k_i) \qquad [6.4]$$

$$k_i = f(k'_i) = f(k_i \oplus e) \qquad [6.5]$$

where e represents the dissimilarity bits between U_i and U_{ip}, and $|e| \leq t$.

The various stages of this phase are shown in Figure 6.4.

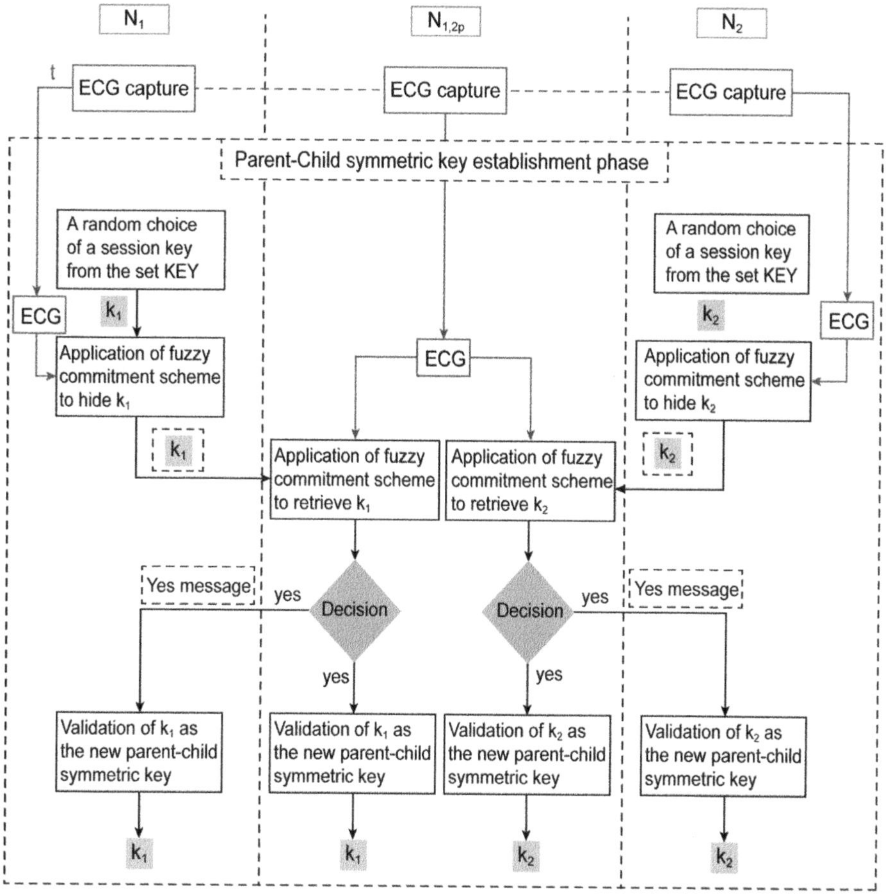

Figure 6.4. *Establishing symmetrical keys between a parent node and its child nodes*

6.4.3.2. *Symmetrical key generation phase using biometric data between sons*

Establishing a biometric session key between two nodes with the same father requires the availability of an ECG signal recorded at the same time for all three nodes (the father and its two sons). This phase consists of three processes (Figure 6.5):

1) generation of a primary key;

2) verification and decision of the father;

3) generation of the biometric session key.

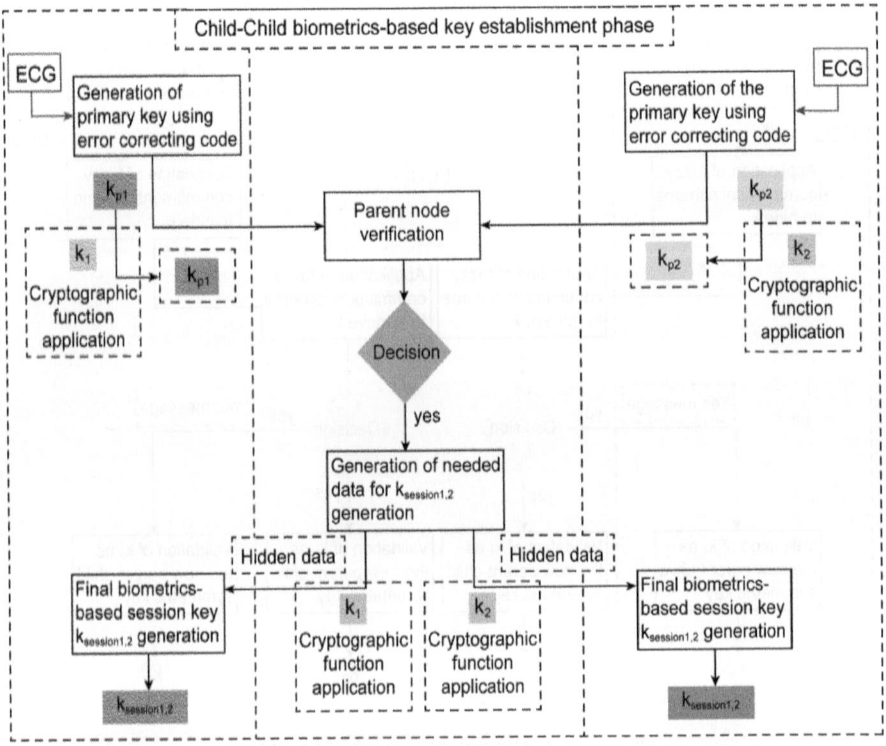

Figure 6.5. *Establishing the symmetrical key between two child nodes*

6.4.3.2.1. Generating a primary key

– Step 1: primary key generation is a two-stage process. In the first step, each child node generates a primary key k_p using an ECC and the result of the binary coding of the measured ECG signal. The use of the ECC enables the correction of dissimilarity bits between ECG signals measured by sensors of the same WBAN. In this approach, U_i is considered as a word derived from the codeword. First, U_i is divided into x n-bit blocks $B_i = \{b^1{}_i, b2i, \ldots, b^x{}_i\}$.

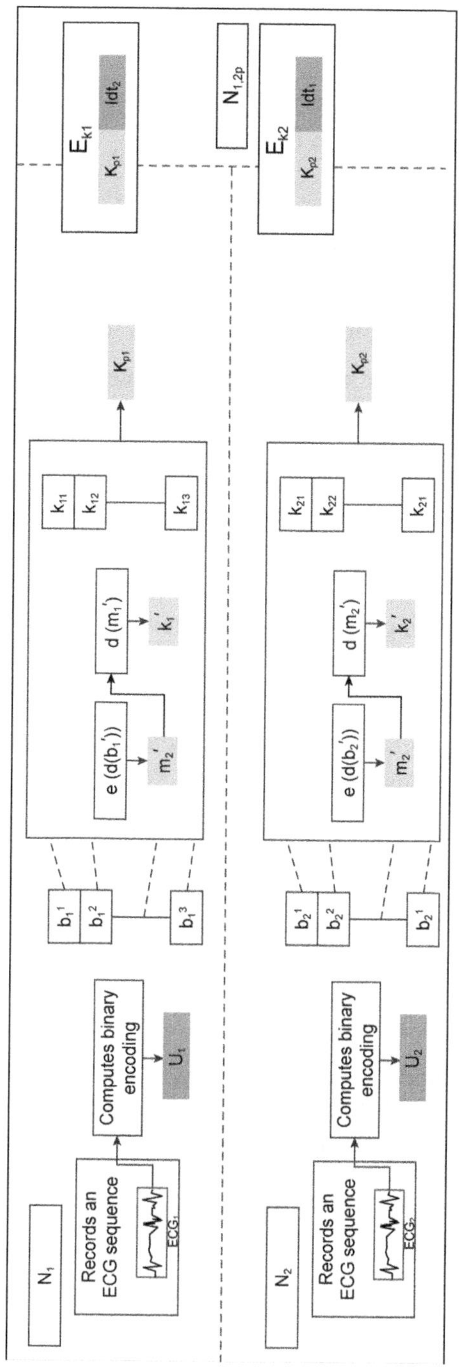

Figure 6.6. *Primary key generation by nodes N_1 and N_2*

Next, for each block, m^l_i is computed:

$$m^l_i = e(d(b^l_i)) \text{ avec } 1 \leq l \leq x \quad [6.6]$$

where m^l_i is the closest codeword to the b^l_i block:

$$k_{li} = d(m^l_i) \text{ avec } 1 \leq l \leq x \quad [6.7]$$

where k_{li} is a part of the primary key.

The k_{pi} primary key is the concatenation of the k_{li} keys.

– Step 2: once the primary keys k_{p1} and k_{p2} have been generated by the two child nodes, each node must transmit the result of the concatenation of its primary key with the Idt of the other node to its father. This message must be protected with the father–child session key generated in the first phase (see Figure 6.6).

6.4.3.2.2. Checking and deciding on the father node

– Step 3: after decrypting the messages received from the two sons, the father calculates the dissimilarity bits (Hamming distance [HD]) between the primary keys received (k_{p1} and k_{p2}).

– Step 4: the father generates a random sequence of bits called m_{index}. This sequence will be used to generate the biometric key.

– Step 5: the father arbitrarily chooses one of the received primary keys and transmits the XOR of the *HD* calculated (HD = $k_{p1} \oplus k_{p2}$) and the generated m_{index} to one of the two child nodes, and the generated m_{index} sequence to the other. Both messages are encrypted with the corresponding father–child session keys.

6.4.3.2.3. Generating a biometric session key

– Step 6: each child node, N_1 and N_2, decrypts the message received from its parent. Then, both nodes must perform the following operations to calculate the same biometric session key $ksession_{12}$ (see Figure 6.7):

$$N_1: k_{session12} = M(k_{p1} \oplus HD \oplus m_{index}) = M(k_{p2} \oplus m_{index}) \quad [6.8]$$

$$N_2: k_{session12} = M(k_{p2} \oplus m_{index}) \quad [6.9]$$

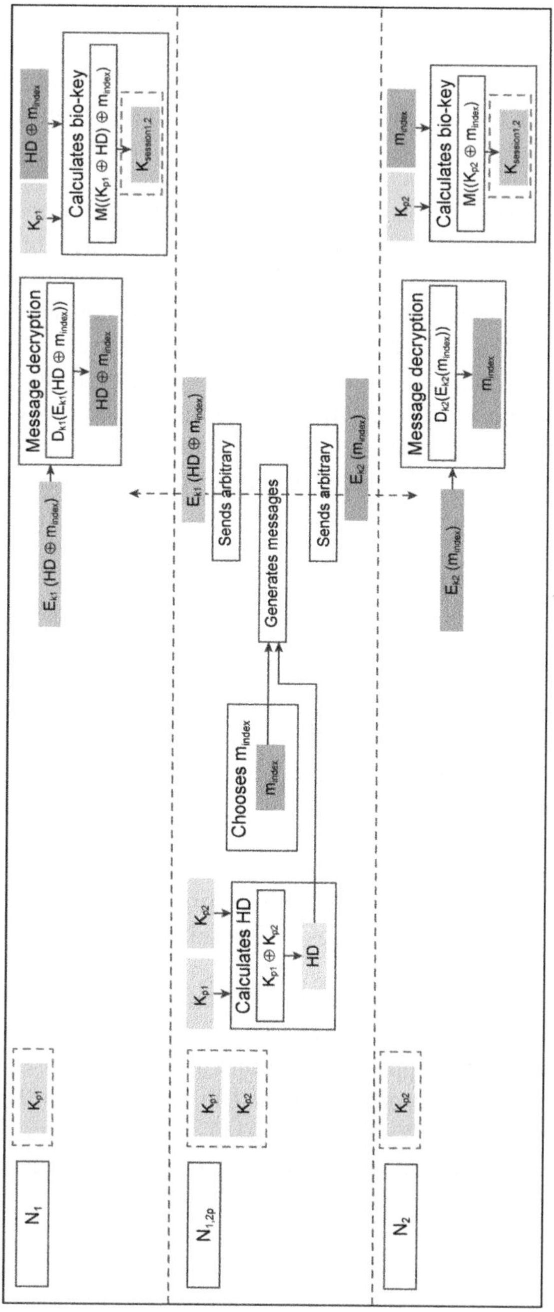

Figure 6.7. *Generating a biometric session key*

In the event that the two nodes wishing to secure their communication do not have the same father node, their closest common ancestor will play the role of the third node. The process of establishing the symmetrical key generated from the ECG is similar in the case of the father node. However, it must be preceded by the discovery of the closest common ancestor by both nodes.

For the protocol to function properly, Sammoud et al. propose that each parent node must be able to verify the membership of a network node to all its descendants via its Idt. So, when the child node transmits the result of the encryption (with the father–child session key) of the concatenation of the primary key it generated with the Idt of the other node, to its father node, the latter must verify the membership of the second node to all its descendants. If so, it will act as the third node. If not, it transfers the message received after decryption and re-encryption (with its father-child session key) to its father node. This transfer is repeated until the nearest common ascending node is found.

6.5. Discussion

The major challenge facing proposals for securing intra-WBAN communications is to guarantee efficient symmetrical key establishment (generation and distribution) between WBAN nodes, while ensuring a balance between resistance to possible attacks and resource consumption (energy, memory and computation).

As Table 6.1 shows, the approaches studied in this chapter do not provide an ideal solution for securing intra-WBAN communications. Indeed, each of these proposals has its own advantages and disadvantages.

The first proposal by Bui et al. (2007) ensures efficient session key sharing between WBAN nodes, as well as minimal resource consumption. However, unlike the other approaches, biometrics is only used for session key sharing. The second approach proposed in Bui et al. (2007) as well as the approaches proposed in Yao et al. (2011) and Zaghouani et al. (2015) have low energy consumption during session key establishment. However, these approaches have a success rate (obtaining an identical session key by both nodes) of less than 100%. For the proposals by Mana et al. (2009), Jammali et al. (2015) and Sammoud et al. (2020), session key distribution is ensured with a success rate of 100%. Nevertheless, these approaches are not optimal in terms of energy consumption. The proposal by Cho et al. (2016) ensures session key generation and sharing with low computational cost. The approach proposed in Cho et al. (2017) guarantees optimal session key establishment time. Yet still, the drawback of both approaches proposed in Cho et al. (2016, 2017) is the session key recovery success rate, which is less than 100%.

Regarding security, the two approaches proposed in Bui et al. (2007), as well as those proposed in Mana et al. (2009) and Cho et al. (2016) have not been evaluated. As for the proposal by Zaghouani et al. (2015), they claimed that their proposal is robust against replay attacks, passive eavesdropping attacks and MITM attacks. Jammali et al. (2015) proved through formal analysis that their approach is protected against replay attacks. Cho et al. (2017) claimed that their proposal is robust against the session key disclosure attack. Furthermore, the security of the approach presented in Sammoud et al. (2020) has been demonstrated by formal and informal analyses, and the authors have shown the resistance of their approach to a good number of attacks, such as replay attacks, identity theft attacks, session key guessing attacks, passive eavesdropping attacks and MITM attacks.

6.6. Conclusion

The data exchanged within WBANs is sensitive and confidential, requiring the deployment of an effective security protocol, while respecting the resource constraints of medical sensors (computation, memory and energy).

In this chapter, we first present the security features of WBANs. Next, we present the various security solutions dedicated to these networks. Subsequently, we focused on biometric data-based approaches, given their interest. Indeed, this type of approach has the advantage of being based on the use of ECG as a biometric method for securing medical sensor exchanges.

Approaches	Biometry used	Year	Advantages	Disadvantages
(Bui et al. 2007, first part)	ECG	2007	– Low resource consumption for reliable session key sharing	– Use of biometrics only for sharing session keys – High resource consumption to ensure ECG availability – No security analysis
(Bui et al. 2007, second part)	ECG	2007	– Low power consumption during session key generation and sharing	– High resource consumption to ensure ECG availability – Key distribution not always efficient (rate < 100%) – No safety analysis studied
(Mana et al. 2009)	ECG	2009	– Reliable session key distribution (100% efficiency rate)	– High resource consumption – No security analysis
(Yao et al. 2011)	ECG	2011	– Low power consumption during session key generation and sharing	– Key distribution not always effective (rate < 100%)
(Zaghouani et al. 2015)	ECG	2015	– Low power consumption during session key generation and sharing – Resistance to replay, eavesdropping and MITM attacks	– Key distribution not always effective (rate < 100%)

Approaches	Biometry used	Year	Advantages	Drawbacks
(Jammali et al. 2015)	ECG	2015	– Reliable session key distribution (100% efficiency rate) – Resistance to replay attacks	– High resource consumption
(Cho et al. 2016)	PPG	2016	– Low computation cost during session key creation	– Key distribution not always effective (rate < 100%)
(Cho et al. 2017)	PPG	2017	– Resistance to leaked session key attacks	– Key distribution not always effective (rate < 100%)
(Sammoud et al. 2020)	ECG	2020	– Reliable session key distribution (100% efficiency rate) – Resistance to replay attacks, identity theft attacks, session key guessing attacks, passive eavesdropping attacks and MITM attacks	– Sub-optimal resource consumption

Table 6.3. *Biometric-based approaches to securing intra-BAN communications*

6.7. References

Al-Janabi, S., Al-Shourbaji, I., Shojafar, M., Shamshirband, S. (2017). Survey of main challenges (security and privacy) in wireless body area networks for healthcare applications. *Egyptian Informatics Journal*, 18(2), 113–122.

Ali, A. and Khan, F.A. (2015). Key agreement schemes in wireless body area networks: Taxonomy and state-of-the-art. *Journal of Medical Systems*, 39(10), 1–14.

An, Y. (2012). Security analysis and enhancements of an effective biometric-based remote user authentication scheme using smart cards. *Journal of Biomedicine and Biotechnology*.

Bui, F.M. and Hatzinakos, D. (2007). Biometric methods for secure communications in body sensor networks: Resource-efficient key management and signal-level data scrambling. *EURASIP Journal on Advances in Signal Processing*, 2008, 1–16.

Charlton, P.H., Kyriacou, P., Mant, J., Alastruey, J. (2020). Acquiring wearable photoplethysmo-graphy data in daily life: The PPG diary pilot study. *Engineering Proceedings*, 2(1), 80.

Cho, K. and Chung, B. (2016). Lightweight biometric key agreement scheme for secure body sensor networks. *International Journal of Communications*, 1.

Cho, K. and Chung, B. (2017). Efficient secret key delivery using heartbeats. *MATEC Web of Conferences*, 125(03006). EDP Sciences, Les Ulis.

Hajar, M.S., Al-Kadri, M.O., Kalutarage, H.K. (2021). A survey on wireless body area networks: Architecture, security challenges and research opportunities. *Computers & Security*, 104, 102211.

Hussain, S.Z. and Kumar, M. (2021). Secured key agreement schemes in wireless body area network–A review. *Indian Journal of Science and Technology*, 14(24), 2005–2033.

Jammali, N. and Fourati, L.C. (2015). PFKA: A physiological feature based key agreement for wireless body area network. In *2015 International Conference on Wireless Networks and Mobile Communications (WINCOM)*. IEEE.

Kompara, M. and Hölbl, M. (2018). Survey on security in intra-body area network communication. *Ad Hoc Networks*, 70, 23–43.

Mana, M., Feham, M., Bensaber, B.A. (2009). SEKEBAN (secure and efficient key exchange for wireless body area network). *International Journal of Advanced Science and Technology*, 12, 45–60.

Masdari, M., Ahmadzadeh, S., Bidaki, M. (2017). Key management in wireless body area network: Challenges and issues. *Journal of Network and Computer Applications*, 91, 36–51.

Morales, L.V., Delgado-Ruiz, D., Rueda, S.J. (2019). Comprehensive security for body area networks: A survey. *Int. J. Netw. Secur*, 21, 342–354.

Narwal, B. and Mohapatra, A.K. (2021). A survey on security and authentication in wireless body area networks. *Journal of Systems Architecture*, 113, 101883.

Pirbhulal, S., Samuel, O.W., Wu, W., Sangaiah, A.K., Li, G. (2019). A joint resource-aware and medical data security framework for wearable healthcare systems. *Future Generation Computer Systems*, 95, 382–391.

Poon, C., Zhang, Y.T., Bao, S.D. (2006). A novel biometrics method to secure wireless body area sensor networks for telemedicine and m-health. *IEEE Communications Magazine*, 44(4), 73–81.

Sammoud, A., Chalouf, M.A., Hamdi, O., Montavont, N., Bouallegue, A. (2020). A new biometrics-based key esta-blishment protocol in WBAN: Energy efficiency and security robustness analysis. *Computers & Security*, 96, 101838.

Yao, L., Liu, B., Wu, G., Yao, K., Wang, J. (2011). A biometric key establishment protocol for body area networks. *International Journal of Distributed Sensor Networks*, 7(1), 282986.

Zaghouani, E.K., Jemai, A., Benzina, A., Attia, R. (2015). ELPA: A new key agreement sche-me based on linear prediction of ECG features for WBAN. In *23rd European Signal Processing Conference (EUSIPCO)*. IEEE.

7

Using Biometrics for Authentication in e-Health Systems

Omessaad HAMDI[1], **Mohamed AYMEN CHALOUF**[2] **and Amal SAMMOUD**[3]

[1] *IEEE, Rennes, France*

[2] *IRISA, Université de Rennes, France*

[3] *MIRA, Aflokkat, Ajaccio, France*

7.1. Introduction

To ensure continuous patient monitoring, e-health systems enable healthcare professionals to remotely access medical servers to consult patient medical information. The use of authentication mechanisms to secure communications between a healthcare professional and a medical server is an obligation to preserve the confidentiality of sensitive data for users of e-health applications (healthcare professionals and patients).

Authentication is therefore one of the essential security services required to protect e-health systems by guaranteeing the confidentiality of medical data and respecting patient privacy. In this chapter, we present authentication mechanisms using biometrics.

The chapter is structured as follows. Section 7.2 introduces e-health systems. Section 7.3 introduces authentication techniques. Section 7.4 presents biometric-based authentication. In section 7.5, we describe a multi-factor authentication protocol suitable for e-health systems. Finally, section 7.6 concludes the chapter.

7.2. e-Health systems

In this section, we present the architecture of an e-health system, before listing some of the attacks that can threaten the security of e-health systems, as well as the security services available to counter these attacks. Section 7.3 will focus on one of these security services, authentication.

7.2.1. *Architecture*

Several architectures have been proposed in the literature (Hamdi et al. 2014). Most works converge on a basic architecture comprising:

– a WBAN (wireless body area network): composed of sensors that measure, collect and transmit the patient's vital signals to the gateway;

– a gateway: transfers the data collected by the sensors to the medical platform;

– a medical platform: the servers that store, process and secure medical data in the cloud.

Most of the architectures are presented in the literature (Krishnamoorthy et al. 2023).

There are a number of possible attacks on e-health systems (Figure 1.3). Below, we describe the best-known attacks:

– One type of attack prevents users (patients and healthcare professionals) from accessing the e-health system and medical servers.

– Eavesdropping attack: the aim of this attack is to read clandestinely exchanged information.

– Password disclosure attack: the aim of this attack is to guess the patient's and/or healthcare professional's password and attempt to connect to the medical server.

– Identity theft attack: this attack consists of impersonating a legitimate user of the e-health system.

– Man-in-the-middle attack: this involves intercepting messages sent between the user (patient or healthcare professional) and the server and replacing them with messages of the user's own.

– Replay attack: this is an attack in which a transmission is repeated by an attacker who has maliciously intercepted the transmission. This is a type of identity theft attack.

7.2.2. *Security services*

The security of e-health systems requires stringént security measures to ensure required care quality. The following are the security requirements that need to be taken into account for the successful acceptance of e-health systems by users:

– Confidentiality: it guarantees the detection of any modification of data exchanged between authorized entities.

– Integrity: it guarantees that an unauthorized party cannot modify the data exchanged.

– Authentication: it guarantees that the entities communicating are legitimate and authentic.

– Anonymity: it guarantees that the user's identity is not revealed by an unauthorized third party.

– Untraceability: it guarantees that the attacker cannot trace the communication back to the user (patient or healthcare professional) or the medical server.

In this chapter, we focus on authentication in e-health systems using biometrics. Before describing this type of authentication method in detail, section 7.3 introduces the various authentication techniques.

7.3. Authentication techniques

Authentication is a process that enables a computer system to verify the legitimacy of an entity's access request to authorize or deny access to resources.

7.3.1. *Authentication factors*

There are three authentication factors as follows:

– Memory: this technique involves using information that only the access requester knows (e.g. password and code).

– Hardware: this technique involves using a unique piece of information that only the access requester possesses (e.g. token and card).

– Physical body: this technique involves using information that characterizes the access requester in a given context (e.g. a biometric characteristic).

Table 7.1 shows the advantages and disadvantages of each authentication factor.

Authentication factors	Advantages	Disadvantages
Hardware	– Easily replaceable in the event of loss or theft	– Can be stolen – Can be corrupted – Can be shared – The same person can be registered with different identities
Memory	– Simple and economical – Easily replaceable if it is lost	– Difficult to memorize – Can be guessed/voiced – Can be shared – The same person can be registered with different identities
Physical body	– Cannot be guessed, forgotten, stolen or shared – High level of security	– Impossible to replace – The same person cannot be registered with different identities

Table 7.1. *Authentication factors*

7.3.2. *Types of authentications*

There are two types of authentication:

– Simple authentication: authentication is based on a single authentication factor (e.g. password and badge).

– Strong authentication or multi-factor authentication: authentication is based on several factors (at least two) to ensure the authentication of the entity requesting access.

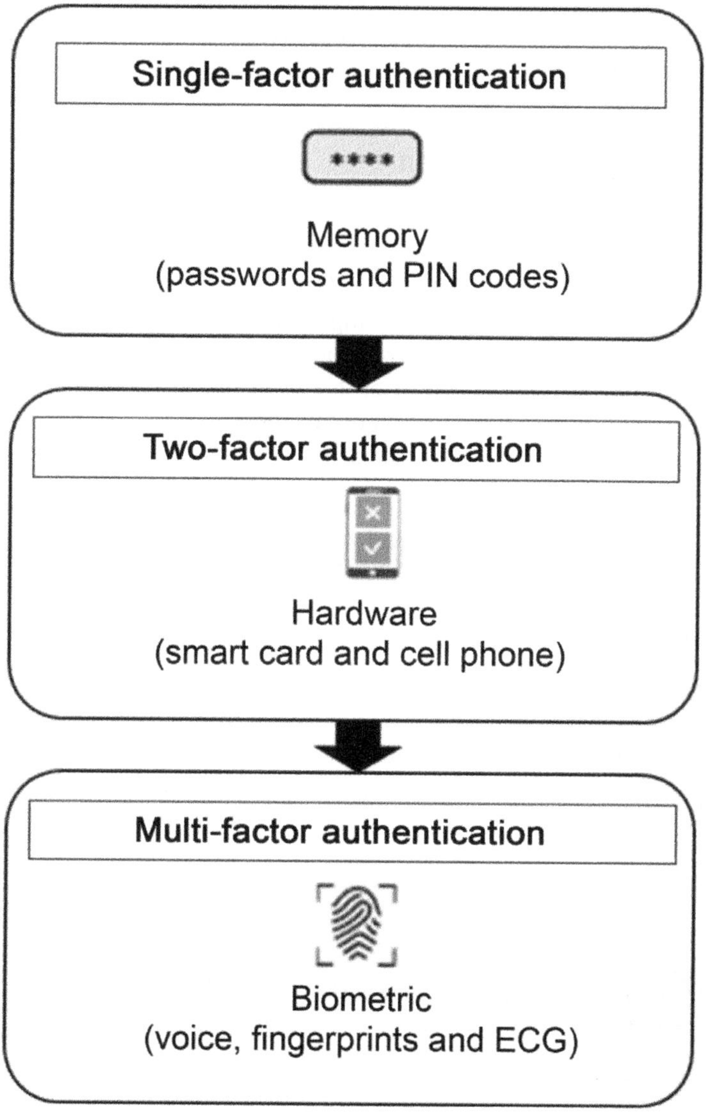

Figure 7.1. *Authentication techniques*

Figure 7.1 illustrates the classic sequence from single to multi-factor authentication. Single-factor authentication is often based on a memory factor, such as a password or PIN. Two-factor authentication often combines a memory factor and a hardware factor, such as a token or magnetic card. Multi-factor authentication uses three factors: a memory factor, a hardware factor and a physical body factor (biometrics). In section 7.4, we look at biometric authentication.

7.4. Biometric authentication

7.4.1. *Biometric features*

Authentication using biometrics (physical body factors) is based on a number of characteristics, such as fingerprint, face, retina and voice. Each characteristic has its own advantages and/or disadvantages and can therefore be adapted to the context of use. Each of these features has its own advantages and/or disadvantages and can therefore be adapted to different contexts of use. Table 7.2 illustrates the advantages and disadvantages of some biometric features.

Modalities	Advantages	Disadvantages
Fingerprint	– Non-invasive – Reliable – Highly acceptable	– Easily forged with an artificial finger – Cut, scar or even absence of finger may be an obstacle
Face	– Non-invasive – Highly acceptable – No physical contact required	– Facial features change over time – Distinctiveness not guaranteed for twins
Retina	– Cannot be tampered with – Reliable – No physical contact required	– Not easy to use – It may be affected by some diseases such as diabetes and cataracts.
Signature	– High acceptability – Non-invasive	– Risk of incoherent signatures
Voice	– Easy to install – No special devices required – Non-invasive	– Noise sensitivity – Easy to hack
ECG	– Cannot be tampered with – Reliable	– Difficult to implement

Table 7.2. *Biometric features*

7.4.2. *Biometric system effectiveness*

The effectiveness of a biometric system (Mason et al. 2020) is measured according to seven criteria:

– Universality: each person has unique characteristics that cannot be reproduced.

– Uniqueness: this criterion refers to the ability of a biometric system to uniquely recognize one user from a group of users.

– Permanence: a registered characteristic must remain constant in the database over a certain period.

– Measurability (or collectability): a characteristic of a person can be easily acquired or measured.

– Performance: this criterion refers to the accuracy of the system in relation to operational requirements.

– Acceptability: this criterion indicates how easy it is for users to accept the technology to facilitate measurement of the biometric characteristic.

– Circumvention: this criterion refers to the ease with which a feature can be imitated by fraudulent means.

Table 7.3 illustrates the level of each criterion for a few biometric features (Fatima et al. 2019).

Criteria	Fingerprint	Face	Retina	Signature	Voice	ECG
Universality	Medium	High	High	Low	Medium	High
Uniqueness	High	Low	High	Low	Low	High
Permanence	High	Medium	Medium	Low	Low	High
Measurability	Medium	High	Low	High	Medium	Medium
Performance	Haute	Low	High	Low	Low	High
Acceptability	Medium	High	Low	High	High	Medium
Bypass	High	Low	High	Low	Low	Low

Table 7.3. *Effectiveness of biometric features (Fatima et al. 2019)*

7.4.3. *Performance measures for biometric systems*

The performance measures of a biometric system are closely related to the false reject ratio (FRR) and false acceptance ratio (FAR) (Fatima et al. 2019):

– FRR indicates the probability that a system will not recognize a biometric feature as coming from the person concerned.

– FAR indicates the probability that a system will incorrectly match a biometric feature from one person to that of another.

7.5. Multimodal authentication

Managing emergency situations in the healthcare sector generally requires immediate action. Despite the urgent nature of these situations, some vital information about the patient's medical history is required before treatment can begin. A delay in accessing the patient's medical records in an emergency situation could have serious effects. The use of approaches (such as tokens or smart cards) to access patients' medical records becomes ineffective in emergency situations, as the patient may be unconscious. In addition, these traditional means of authentication can be lost, falsified, forgotten or stolen. Again, the disadvantages associated with biometric access control based on a single feature for records (such as spoofing or interference attacks) reduce the effectiveness of systems developed with a single biometric modality (Jain et al. 2004; Soleymani et al. 2018; Ogbodo et al. 2020).

Biometric systems can be classified into two categories:

– a unimodal biometric system based on a single source of biometric data is called a unimodal system;

– a multimodal biometric system based on a combination of two or more different biometric sources of an individual is called a multimodal biometric system (Talreja et al. 2017).

Multimodal biometrics is an interesting research topic. Several approaches have been proposed using multimodal biometric systems (Barbu et al. 2015).

Cherrat et al. (2020) illustrate a multimodal biometric system based on voice and facial authentication. In Schiavone et al. (2016), a multimodal biometric authentication system that verifies continuously the user's identity using face, fingerprint and keystroke has been proposed. In Byahatti et al. (2020) and Aldija et al. (2021), three biometric sources were used to design a multimodal system: face, iris and fingerprint.

In Barra et al. 2017), a multimodal system using electrocardiogram (ECG) and electroencephalogram (EEG) was proposed.

Table 7.4 summarizes some proposed multimodal biometric authentication schemes.

References	Biometric data
(Cherrat et al. 2020)	Voice + face
(Schiavone et al. 2016)	Face + fingerprint + keyboard typing
(Byahatti et al. 2020; Aldjia et al. 2021)	Face + iris + fingerprint
(Barra et al. 2017)	ECG + EGG

Table 7.4. *Multimodal authentication*

7.6. Multi-factor authentication approaches for e-health system security

Several multifactor authentication approaches have been proposed to secure communications in e-health applications, specifically, communications between the healthcare professional and the medical server.

An (2012) has proposed an authentication approach based on three factors: memory, hardware and biometrics. The proposed approach aims to correct the security flaws of the approach introduced in Das et al. (2011). However, study of the approach in An (2012), presented in the research work of Khan et al. (2013), Cao et al. (2015), Das et al. (2015) and Li et al. (2016), shows that it has security flaws and low resistance to password disclosure and identity theft attacks (user and server).

To address these security flaws, Khan et al. (2013) proposed an improvement on the approach of An (2012), also based on three authentication factors. They asserted the robustness of their proposal against a number of attacks, such as password disclosure and identity theft attacks. However, in 2015, Wen et al. (2015) demonstrated that the proposal of Khan et al. (2013) does not guarantee the security requirements mentioned. They therefore proposed an improved approach to Khan et al. (2013). However, this proposal has a significant cost in terms of resource consumption (computation, communication and memory). In addition, this approach does not guarantee user anonymity. In 2020, Sammoud et al. (2020) proposed an improvement on the approach of Khan et al. (2013). This new approach is also based on three factors (memory, hardware and biometrics) to ensure mutual authentication between the healthcare professional and the medical server. Based on formal and informal security analyses, Sammoud et al. (2020) confirmed the resistance of their proposal to the attacks considered and the correction of security flaws in the approach of Khan et al. (2013), while protecting user anonymity and minimizing resource consumption.

Similar to the approach of Khan et al. (2013), Cao et al. (2015), Das et al. (2015) and Li et al. (2016) have all proposed an improvement to the protocol of An (2012) based on three authentication factors in order to overcome its security issues. However, the approach of Das et al. (2015) has a significant cost in terms of computation and communications. The approach of Cao et al. (2015) is not optimal in terms of communications and does not ensure session key establishment between user and server. In Li et al. (2016), establishing efficient authentication with session key generation between the healthcare professional and the medical server has a fairly high cost in terms of computation and communications.

In Chaudhry et al. (2018), a multi-factor authentication approach has been proposed to secure communications between the healthcare professional and the medical server, while correcting the security flaws in Mir et al.'s (2015) proposal related to anonymity and impersonation (user and server). The security analysis provided in Chaudhry et al. (2018) shows that this proposal corrects the security issues of Mir et al.'s (2015) approach.

Al-Saggaf et al. (2019) have proposed a three-factor authentication protocol for healthcare professional access to the medical server. The security analysis of this protocol shows its resistance to attacks, such as impersonation of the healthcare professional, as well as the medical server, password disclosure and internal attacks.

Approaches	Computation operations	Resource consumption (computing, communication and memory)
(An 2012)	– Hash function	– Low consumption of smart card computing, communications and memory resources
(Khan et al. 2013)	– Hash function	– Low consumption of smart card computing, communications and memory resources
(Wen et al. 2015)	– Hash function – Fuzzy extractors	– High consumption of smart card computing, communications and memory resources
(Das et al. 2015)	– Hash function – Encryption/decryption functions	– Low consumption of smart card memory resources – High resource consumption in terms of computations and communications
(Cao et al. 2015)	– Hash function	– Low consumption of smart card memory and computing resources – High consumption of communication resources
(Li et al. 2016)	– Hash function – Fuzzy extractors – Elliptic curve cryptography	– Low resource consumption in terms of communications and smart card memory – High consumption of computing resources
(Chaudhry et al. 2018)	– Hash function – Encryption/decryption functions	– Low consumption of smart card memory resources – High consumption of computing and communications resources
(Al-Saggaf et al. 2019)	– Hash function – Encryption/decryption functions	– High consumption of smart card computing, communications and memory resources
(Sammoud et al. 2020)	– Hash function – Fuzzy extractors	– Low consumption of smart card memory and communications resources – High consumption of computing resources

Table 7.5. *Comparison of resource consumption of multifactor authentication approaches between the healthcare professional and the medical server*

Approaches	Security analysis		A1	A2	A3	A4	A5	A6	A7	A8
	Formal	Informal								
(An 2012)	Yes	No	No	No	No	NA	No	No	No	No
(Khan et al. 2013)	Yes	No	No	No	No	NA	No	No	No	No
(Wen et al. 2015)	Yes	BAN logic	Yes	Yes	Yes	Yes	NA	No	Yes	Yes
(Das et al. 2015)	Yes	AVISPA	Yes	NA	NA	Yes	NA	Yes	Yes	Yes
(Cao et al. 2015)	Yes	BAN logic	Yes	Yes	NA	NA	Yes	NA	NA	No
(Li et al. 2016)	Yes	No	Yes	NA	Yes	Yes	Yes	NA	Yes	Yes
(Chaudhry et al. 2018)	Yes	ProVerif	Yes	Yes	Yes	Yes	NA	Yes	Yes	Yes
(Al-Saggaf et al. 2019)	Yes	No	NA	NA	NA	Yes	NA	Yes	NA	Yes
(Sammoud et al. 2020)	Yes	BAN Logic + AVISPA	Yes	Yes	Yes	Yes	Yes	Yes	Yes	Yes

A1: resistance to user impersonation attacks. A2: resistance to server identity theft attacks. A3: resistance to password disclosure attacks. A4: resistance to internal attacks. A5: mutual authentication. A6: protection of user anonymity. A7: resistance to replay attacks. A8: establishment of a session key between user and server. NA: information not available.

Table 7.6. *Security comparison of multifactor authentication approaches between the healthcare professional and the medical server*

Table 7.5 illustrates, for each of the proposals studied in the literature, the performance in terms of computing operations, communications and memory required for storage on the smart card. As Table 7.5 shows, the multifactor authentication approaches described in this section are not optimal in terms of resources consumed. This is mainly due to the additional operations used to enhance security. Indeed, the main objective of this research work is to simultaneously ensure acceptable resource consumption (calculations, communications and smart card memory), with a good level of robustness against the best-known attacks. Table 7.6 compares the various multi-factor authentication approaches between healthcare professionals and a medical server in terms of resistance to the attacks considered, such as identity theft and password disclosure attacks. However, according to Table 7.6, with the exception of Sammoud et al.'s (2020) approach, whose resistance to the attacks considered has been proven with formal and informal analyses, all the other approaches have security flaws. We can therefore consider that the Sammoud et al.'s (2020) approach offers the best compromise between resource consumption and security robustness when compared with the other approaches introduced in this chapter.

7.6.1. *Sammoud* et al.'s *approach*

In this section, we describe our approach. The proposed approach ensures multi-factor authentication between the healthcare professional (user) and the medical server. This authentication is based on three factors: a memory factor, which is the user's password; a hardware factor, which is the user's smart card; and a physical body factor, which is the user's fingerprint. In addition, this proposal enables the generation of session keys, based on biometric data (the user's ECG), for secure communications between these two entities (the healthcare professional's equipment and the medical server). Following formal and informal security analyses, the authors demonstrated that this protocol corrects the security flaws identified in the approach of Khan et al. (2013), while minimizing resource consumption (calculations, communications and smart card memory).

The operation of this protocol is composed of four phases: registration, connection, authentication and password change, and involves three participants: the healthcare professional (the user) (U_i), the medical server (S) and the registration center (R) representing the trusted entity.

The notations used in the description of Sammoud et al.'s (2020) approach are summarized in Table 7.7.

Notations	Explanations
ID_i	User ID (healthcare professional) U_i
ID_s	Server identifier S
B_i	Biometric characteristics of user U_i (fingerprint)
PW_i	User U_i password
K_i	Random number generated by user U_i
xs	Secret key held by server S
ys	Secret key held by server S
SK	Symmetric key generated to secure communications between U_i and S
Gen, Rep	Probabilistic generation procedure and deterministic reproduction procedure of fuzzy extractor
h (.)	Hash function
a || b	Concatenation between two binary sequences a and b
a ⊕ b	XOR function of two binary sequences a and b

Table 7.7. *Notations used in the description of Sammoud et al.'s (2020) approach*

7.6.1.1. *User registration phase*

This phase is necessary to enable a new user to register with a registration center R in order to be authorized to access the resources of the medical server S. User registration takes place in several stages (Figure 7.2):

– Step 1: U_i chooses an identity ID_i and a password PW_i. Then, they print their biometric data B_i (which is the fingerprint) on a specific device. Using the fuzzy extractor function, the system calculates $(B^*_i, P_i) = Gen(B_i)$. Note that B^*_i represents the biometric key and P_i is the public reproduction parameter that will be used later to reproduce B^*_i. The purpose of using the fuzzy extractor is to ensure the correct reproduction of identical biometric data. After this, U_i transmits its password and masked biometric data ($PW_i \oplus B^*_i$) and its ID_i identity to R via a secure channel.

– Step 2: after receiving U_i's message, R first generates a random number R_n that is unique for each user. Using the server's secret key xs, as well as its identity ID_s, R calculates $f_i = h(PW_i \oplus B^*_i)$, ri = h($ID_i$ ||

$(PW_i \oplus B^*_i)) \oplus f_i$, $c_i = h(ID_s \,||\, xs \,||\, R_n) \oplus f_i$, $e_i = h(ID_i \,||\, xs) \oplus r_i$ et, $g_i = ID_i \oplus (PW_i \oplus B^*_i) \oplus f_i$. Next, it stores the parameters $\{c_i, e_i, g_i, h(.)\}$ in the SC_i smart card before sending it to the user U_i via a secure channel (e.g. registered or hand-delivered mail).

– Step 3: after receiving their smart card, U_i stores P_i in SC_i. So, the parameters stored in SC_i are $\{c_i, e_i, g_i, P_i, h(.)\}$.

Figure 7.2 summarizes the registration phase.

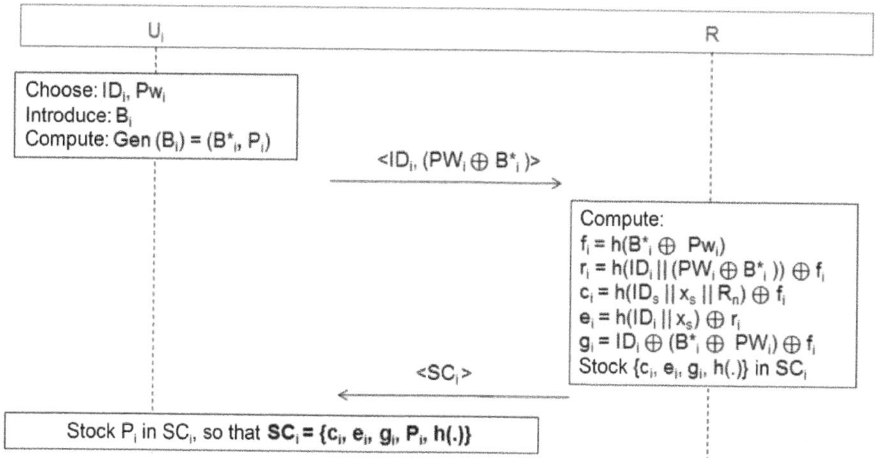

Figure 7.2. *Protocol registration phase*

7.6.1.2. *Connection phase*

To connect to the medical server S, the user U_i, using their SC_i smart card, performs the following steps (Figure 7.3):

– Step 4: after inserting their smart card into the dedicated card reader, U_i enters their identity ID_i and password PW_i. They then print out their biometric data by placing their thumb on the specific device, which simultaneously captures their ECG_i and their B'_i fingerprint. The ECG_i will later be used to generate a random number specific to U_i (R_c). B'_i will be used to verify the user's identity.

– Step 5: using the fuzzy extractor function Rep(.,.), the SC_i smart card regenerates B^*_i, by calculating : $B^*_i = Rep(B'_{i}, _{Pi})$. Next, SC_i calculates $f_i = ID_i \oplus (PW_i \oplus B^*_i) \oplus g_i$. In the event of equality between f_i, stored in the smart card, and h ($B^*_i \oplus PW_i$), user U_i passes their identity check and SC_i proceeds with the operations required for the connection phase. Otherwise, SC_i terminates the connection.

– Step 6: SC_i calculates $r_i = h(ID_i \,||\, (PW_i \oplus B^*_i)) \oplus f_i$, $M1 = c_i \oplus f_i$, $M2 = e_i \oplus r_i$, $M3 = M1 \oplus R_c$, $M4 = M1 \oplus (R_c \oplus ID_i)$ and $M5 = h(M2 \,||\, R_c)$, where R_c is the random number generated from the measured ECG_i. Indeed, thanks to its fairly random nature, ECG is a good alternative to conventional methods based on cryptographic mechanisms or random generators for generating random numbers.

– Step 7: the SC_i smart card then sends the connection request {M3, M4, M5} to the S server via an unsecured channel (e.g. the Internet). Messages M3 and M4 are used by S to verify the identity of user U_i and to retrieve the random number R_c required to generate the symmetric key (SK). M5 is used by the server to verify the integrity of the connection request.

7.6.1.3. *Authentication phase*

To ensure mutual authentication and efficient generation of the session key that will later be used to secure communication between U_i and S, the following steps must be carried out (Figure 7.3):

– Step 8: on receiving the connection request {M3, M4, M5} from the user U_i, the server S checks the identity sent by U_i. It calculates $M6 = h(ID_s||xs||R_n)$, $M7 = M3 \oplus M6$ and $ID^*_i = M4 \oplus (M6 \oplus M7)$. If S finds the calculated identity, stored in its database, then the user's identity is verified and U_i is authenticated by S. Otherwise, S terminates U_i's connection.

– Step 9: once U_i's identity has been validated, S verifies the integrity of message M5. It calculates $M8 = h(ID_i|| \, xs)$ and compares the message M5 received with h ($M8||M7$). S generates a random number R_s to calculate: $M9 = M8 \oplus R_s$ and $M10 = h(M8||R_s)$. Otherwise, S terminates U_i's connection and authentication fails.

– Step 10: S generates the session key: $SK = h(M8|| \, R_s) = h(R_c|| \, R_s)$ and transmits the response message {M9, M10} to user U_i.

– Step 11: on receiving the message {M9, M10} transmitted by S, user U_i, via their smart card SC_i, calculates: $M11 = M9 \oplus M2$ and checks the equality of message M10 with $h(M2||M11)$. In the event of equality, U_i, via its smart card SC_i, generates the same session key $SK = h(R_c \,||\, M11) = h(R_c \,||\, R_s)$. Otherwise, login fails.

Figure 7.3 summarizes the connection and authentication phase of the protocol.

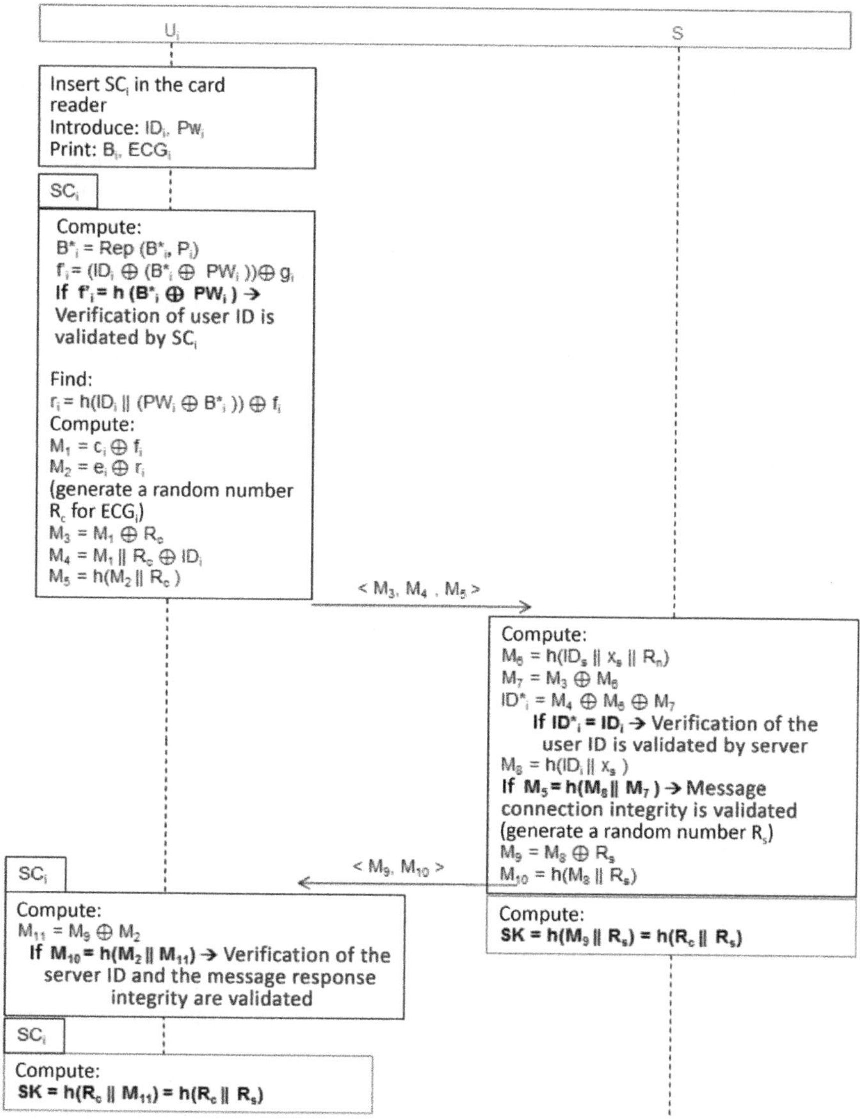

Figure 7.3. *Connection and protocol authentication phase*

7.6.1.4. *Change password phase*

This phase is necessary when a legal user wishes to update their password for security reasons:

– Step 1: after inserting their SC_i smart card into a card reader, U_i then enters their ID_i and password PW_i. They also print their fingerprint B'_i using the specific device.

– Step 2: this step is similar to step 5 of the login phase. First, SC_i calculates $B^*_i = Rep(B'_i, P_i)$ using the stored biometric data B'_i and P_i. Next, it calculates $f^*_i = ID_i \oplus (PW_i \oplus B^*_i) \oplus g_i$. If f^*_i equals $h(B^*_i \oplus PW_i)$, then user U_i's identity is verified, and SC_i asks the user to enter the new password. Otherwise, SC_i rejects the password change request.

– Step 3: using the new password $PW_{i(new)}$, SC_i recalculates the saved parameters: c_i, e_i and g_i. Therefore, it calculates $f_{i(new)} = h(PW_{i(new)} \oplus B^*_i)$, $r_{i(new)} = h(ID_i || (PW_{i(new)} \oplus B^*_i)) \oplus f_{i(new)}$, $c_{i(new)} = h(ID_s || xs || R_n) \oplus f_{i(new)}$, $e_{i(new)} = h(ID_i || xs) \oplus r_{i(new)}$ and $g_{i(new)} = ID_i \oplus (PW_{i(new)} \oplus B^*_i) \oplus f_{i(new)}$. Finally, the SCi smart card updates the following parameters: $c_{i(new)}$, $e_{i(new)}$, $g_{i(new)}$, P_i, $h(.)$.

7.7. Conclusion

In this chapter, we introduced authentication techniques and their uses in e-health systems. We then compared several techniques in terms of security properties, but also in terms of resource consumption. This comparison shows the relevance of Sammoud et al.'s (2020) approach we have detailed in this chapter. This is a mutual authentication approach between a healthcare professional and a medical server. This protocol uses three factors to authenticate the user (healthcare professional): a password, a smart card and a fingerprint. It also uses another user biometric (ECG) to generate and share a symmetrical session key between the two communication endpoints (user equipment and medical server). This protocol ensures a high level of security, while minimizing resource consumption (computation, smart card memory and communications).

As this protocol requires additional equipment to measure the ECG needed to generate the session key, we will investigate the possibility of generating this key based on the basic biometric data (fingerprint) without causing new vulnerabilities and additional costs in terms of resource consumption. This will dispense with the main constraint of this approach: the need for an ECG sensor.

7.8. References

Al-Saggaf, A.A. and Sheltami, T.R. (2019). Renewable and anonymous biometrics-based remote user authentication scheme using smart cards for telecare medicine information system. In *2019 Advances in Science and Engineering Technology International Conferences (ASET)*. IEEE.

Aldjia, B. and Leila, B. (2021). Sensor level fusion for multi-modal biometric identification using deep learning. In *2021 International Conference on Recent Advances in Mathematics and Informatics (ICRAMI)*. IEEE.

An, Y. (2012). Security analysis and enhancements of an effective biometric-based remote user authentication scheme using smart cards. *BioMed Research International*, 1, 519723.

Barbu, T., Ciobanu, A., Luca, M. (2015). Multimodal biometric authentication based on voice, face and iris. In *2015 E-Health and Bioengineering Conference (EHB)*. IEEE.

Barra, S., Casanova, A., Fraschini, M., Nappi, M. (2017). Fusion of physiological measures for multimodal biometric systems. *Multimedia Tools and Applications*, 76, 4835–4847.

Byahatti, P. and Shettar, M.S. (2020). Fusion strategies for multimodal biometric system using face and voice cues. In *IOP Conference Series: Materials Science and Engineering*, 925, 012031.

Cao, L. and Ge, W. (2015). Analysis and improvement of a multi-factor biometric authentication scheme. *Security and Communication Networks*, 8(4), 617–625.

Chaudhry, S.A., Naqvi, H., Khan, M.K. (2018). An enhanced lightweight anonymous biometric based authentication scheme for TMIS. *Multimedia Tools and Applications*, 77, 5503–5524.

Cherrat, M., El Alaoui, R., Bouzahir, H. (2020). Convolutional neural networks approach for multimodal biometric identification system using the fusion of fingerprint, finger-vein and face images. *PeerJ Computer Science*, 6, e248.

Das, A.K. (2011). Analysis and improvement on an efficient biometric-based remote user authentication scheme using smart cards. *IET Information Security*, 5(3), 145–151.

Das, A.K. and Goswami, A. (2015). A robust anonymous biometric-based remote user authentication scheme using smart cards. *Journal of King Saud University-Computer and Information Sciences*, 27(2), 193–210.

Fatima, K., Nawaz, S., Mehrban, S. (2019). Biometric authentication in health care sector: A survey. In *2019 International Conference on Innovative Computing (ICIC)*. IEEE.

Hajar, M.S., Al-Kadri, M.O., Kalutarage, H.K. (2021). A survey on wireless body area networks: Architecture, security challenges and research opportunities. *Computers & Security*, 104, 102211.

Hamdi, O., Chalouf, M.A., Ouattara, D., Krief, F. (2014). e-Health: Survey on research projects, comparative study of telemonitoring architectures and main issues. *Journal of Network and Computer Applications*, 46, 100–112.

Jain, A.K., Ross, A., Prabhakar, S. (2004). An introduction to biometric recognition. *IEEE Transactions on Circuits and Systems for Video Technology*, 14(1), 4–20.

Khan, M.K. and Kumari, S. (2013). An improved biometrics-based remote user authentication scheme with user anonymity. *BioMed Research International*, 1, 491289.

Li, X., Niu, J., Khan, M.K., Liao, J., Zhao, X. (2016). Robust three-factor remote user authentication scheme with key agreement for multimedia systems. *Security and Communication Networks*, 9(13), 1916–1927.

Mason, J., Dave, R., Chatterjee, P., Graham-Allen, I., Esterline, A., Roy, K. (2020). An investigation of biometric authentication in the healthcare environment. *Array*, 8, 100042.

Mir, O. and Nikooghadam, M. (2015). A secure biometrics based authentication with key agreement scheme in telemedicine networks for e-health services. *Wireless Personal Communications*, 83, 2439–2461.

Ogbodo, I.A. (2020). Exploring access to EHR by emergency patients using multimodal biometrics. *Int. J. Latest Technol. Eng. Manag. Appl. Sci*, 9.

Sammoud, A., Chalouf, M.A., Hamdi, O., Montavont, N., Bouallegue, A. (2020). A secure three-factor authentication and biometrics-based key agreement scheme for TMIS with user anonymity. In *International Wireless Communications and Mobile Computing (IWCMC)*. IEEE.

Schiavone, E., Ceccarelli, A., Bondavalli, A., Carvalho, A. (2016). Usability assessment in a multi-biometric continuous authentication system. In *2016 Seventh Latin-American Symposium on Dependable Computing (LADC)*. IEEE.

Soleymani, S., Dabouei, A., Kazemi, H., Dawson, J., Nasrabadi, N.M. (2018). Multi-level feature abstraction from convolutional neural networks for multimodal biometric identification. In *24th International Conference on Pattern Recognition (ICPR)*. IEEE.

Talreja, V., Valenti, M.C., Nasrabadi, N.M. (2017). Multibiometric secure system based on deep learning. In *2017 IEEE Global Conference on Signal and Information Processing (GlobalSIP)*.

Wen, F., Susilo, W., Yang, G. (2015). Analysis and improvement on a biometric-based remote user authentication scheme using smart cards. *Wireless Personal Communications*, 80(4), 1747–1760.

8

Security of Medical Data Processing

Manel ABDELHEDI[1] **and Omessaad HAMDI**[2]
[1] *Noccs, ENISo, Université de Sousse, Tunisia*
[2] *IEEE, Rennes, France*

8.1. Introduction

Recent advances in information and communication technologies (ICT) have undoubtedly improved services in every sector of the world. In particular, information technology has led to vital innovation in the healthcare sector. This has been clearly seen in the transformation from paper-based to electronic medical records, particularly in most of the world's developed countries. The need to combine and integrate a variety of electronic health information from different domains, such as medical research laboratories, hospitals, and health insurance companies, has led to the evolution of a concept known as electronic health (e-health). It should be noted that the different domains involved in sharing medical data have made the application very difficult to manage, hence the need for a cloud-based computing environment for collaborative information sharing between several domains.

Cloud computing is a distributed network that provides services from a provider to users in a number of areas, and in particular in healthcare. This introduction of cloud services into the healthcare sector has several advantages. First, it facilitates the sharing of individuals' electronic medical records between hospitals, as well as to researchers around the world, which

e-Health Security Management,
coordinated by Omessaad HAMDI.

can accelerate progress in medical research. Second, it enables medical records to be stored. Third, users can outsource their complex calculations to a remote server and benefit from the powers of low-cost cloud computing.

However, e-health data contain sensitive and strictly confidential information about patients, such as medical history, demographic data, medications, vaccination status, laboratory test reports, social security number and credit card details. This data can also be consulted at any time. In addition, these data can be accessed by various entities, such as doctors, nurses, pharmacists, laboratory personnel, dentists and medical insurance personnel. Storing or processing these data on third-party servers, which may be unreliable, increases its vulnerability.

So, the main challenge is to protect the confidentiality of sensitive data. Consequently, the user must encrypt the data before storing it on the remote cloud server. However, if a data owner asks the cloud to perform operations on encrypted data, they must share their private key with the remote cloud server. This means that traditional encryption methods (symmetrical or asymmetrical) are becoming insufficient to secure data in the cloud.

Recently, modern cryptographic methods have been proposed to securely externalize data computation in the cloud, homomorphic encryption (HE), and to control access to encrypted data, attribute-based encryption (ABE).

HE enables us to perform calculations on encrypted data without needing to access the plain data. The concept was first introduced by Rivest et al. (1978), just after the invention of RSA. Over the following years, several public-key cryptosystems were proposed, enabling either addition or multiplication to be performed on encrypted data. This family of schemes is known as partially homomorphic encryption (PHE). The best-known PHE encryption schemes are Goldwasser and Micali (GM) (Goldwasser and Micali 1982), ElGamel (ElGamal 1985) and Paillier (Paillier 1999), etc. The first fully homomorphic encryption (FHE) scheme to evaluate any type of function on encrypted data was proposed in 2009 by Gentry (2009a).

ABE is used to control access to encrypted data. It was introduced by Sahai and Waters (2005). It is based on public-key encryption to protect data in the cloud, where encryption and decryption are based on user attributes. In ABE, encryption is based on an access policy in which ciphertext can only be decrypted when the user's attributes match the attributes of the ciphertext.

The aim of this chapter is to study the different families of HE and the different ABE schemes. We also look at how HE and ABE can help secure e-health solutions in cloud computing.

This chapter is structured as follows. Section 8.2 describes the principle of HE, its main families and a comparative study of some of its schemes. The end of this section presents the application of HE to secure e-health solutions. Section 8.3 examines ABE and presents its contribution to securing e-health solutions. Finally, section 8.4 concludes this chapter.

8.2. Homomorphic encryption

8.2.1. *Definition*

HE protects data processing. HE enables mathematical operations to be carried out on encrypted data without needing to access the clear data, giving the same result as if the calculations had been carried out on clear data. This technique has many applications, particularly in cloud computing. Calculations on sensitive encrypted data, such as medical data, can be delegated to a remote server, without compromising confidentiality.

The notion of homomorphism was first introduced by Rivest et al. (1978). This property can translate an operation on encrypted data into an operation on the corresponding plaintext data. Figure 8.1 illustrates the concept of HE for the case of e-health solutions in cloud computing.

From Figure 8.1, we can see that the user (patient, hospital or doctor) must encrypt the patient data (symptoms, etc.) with the patient's public key and send it to the cloud provider. With traditional encryption methods, if the data owner asks the cloud to perform the calculation, they must share their private key with the cloud provider, in order to decrypt the data and calculate the required operations. This makes traditional encryption methods inadequate for securing data in the cloud. With HE, the cloud provider can calculate the operations on the encrypted data, and then return the result without decryption. Finally, the user with the private key (patient, authorized doctor) decrypts the result and obtains the required operation as if the calculation were made on the raw data.

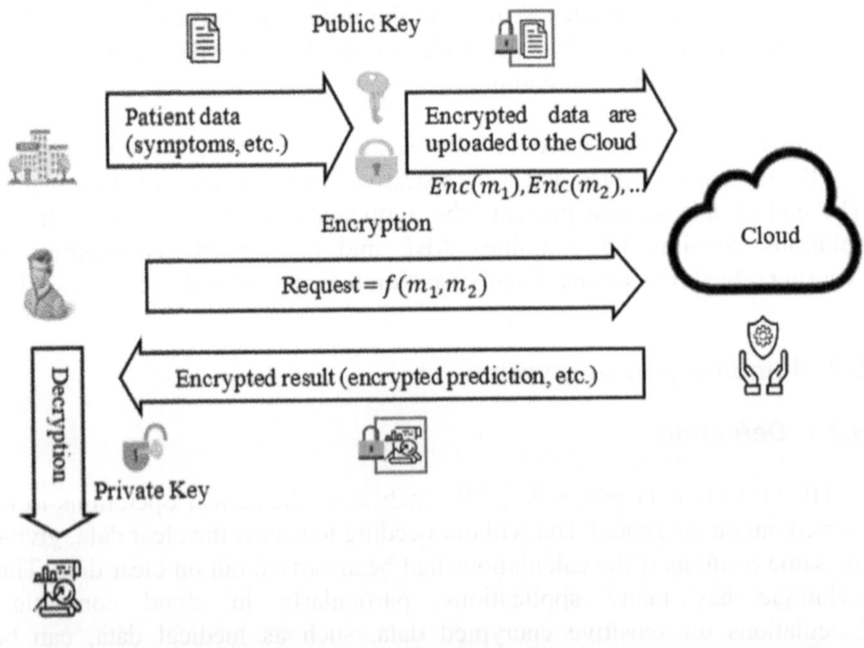

Figure 8.1. *Homomorphic encryption for e-health solutions in cloud computing*

8.2.2. *Terminology*

A homomorphic scheme is defined by four algorithms:

– Key generation KeyGen (λ): in an asymmetric encryption scheme, this algorithm takes the security parameter as input λ and returns the keys (p_k, s_k, e_k), where p_k is the public key, s_k represents the private key and e_k is the evaluation key. In the symmetric encryption scheme, it takes the security parameter λ and returns the keys (s_k, e_k), where s_k is the private key and e_k is the evaluation key.

– Encryption $\mathrm{Enc}(\mathrm{Key}_1, m)$: the encryption algorithm generates the ciphertext c. Its input is the plaintext message m and the key Key_1. This key represents p_k for the asymmetric encryption scheme and s_k for the symmetric encryption scheme.

– Decryption $\mathrm{Dec}(\mathrm{Key}_2, c)$: the decryption algorithm returns the plaintext message. Its input parameters are the cipher c and the key Key_2.

This key is p_k for the asymmetric encryption scheme and s_k for the symmetric encryption scheme.

– Evaluation algorithm Eval($e_k, f, c_1 \dots c_l$): the evaluation algorithm takes as input the evaluation key e_k, a function $f : \{0,1\}^l \longrightarrow \{0,1\}$ and a set of ciphertexts $c_1, c_2, \dots, c_l$. The evaluation algorithm returns a ciphertext c_f, which can be decrypted into $f(m_1, \dots, m_l)$.

The set $\mathcal{F}$ of functions $f \in \mathcal{F}$ determines the homomorphic capabilities of a schema.

Figure 8.2 shows the four algorithms of a HE scheme with asymmetric key. We note that a generic method has been introduced in Rothblum (2011) for converting homomorphic private-key encryption schemes into homomorphic public-key schemes, and vice versa.

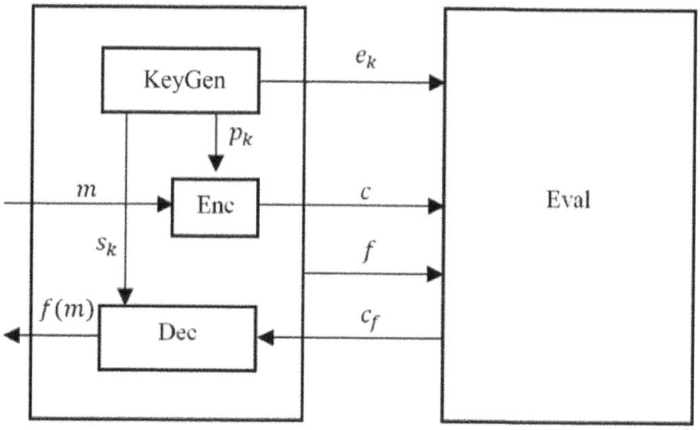

Figure 8.2. *Stages of an asymmetric homomorphic encryption scheme*

In the following, we describe the three main classes of homomorphic schemes. This classification is made according to the set $\mathcal{F}$ of functions it can evaluate.

8.2.3. *Partially homomorphic encryption*

PHE enables a single operation – either addition or multiplication, but not both – to be performed in any number.

PHE schemes include additive HE and multiplicative HE.

8.2.3.1. *Multiplicative homomorphic encryption*

One scheme (KeyGen, Enc, Dec, Eval), based on multiplicative HE, allows only multiplications to be performed on clear data. For this type of HE, there is an operation that verifies equation [8.1]:

$$Enc(p_k, m_1) \cdot \ldots \cdot Enc(p_k, m_l) = Enc(p_k, m_1 \times \ldots \times m_l) \qquad [8.1]$$

Multiplicative homomorphic schemes include the RSA cryptosystem (Rivest et al. 1978) and the ElGamel cryptosystem (1985).

In the following, we use the RSA scheme as an example to explain the principle of a multiplicative HE schemes. This scheme is the first PHE scheme proposed by Rivest et al. (1978) in 1978. RSA is described as follows:

– KeyGen(λ): first, compute $n = pq$ and $\phi = (p-1)(q-1)$, where p and q are two distinct large prime numbers. Then choose e such that $1 < e < \phi$ and $gcd(e, \phi). = 1$. Then, calculate the decryption key d such that $ed \equiv 1 \bmod \phi$. Finally, the public key is $p_k = (e, n)$ and the private key is $s_k = d$.

– Enc(p_k, m) encryption: first, the message is transformed into a plaintext m such that $0 \leq m < n$, then the ciphertext c is calculated by the RSA encryption algorithm as follows:

$$c = m^e \bmod n, \forall\, m \in M \qquad [8.2]$$

where M is the set of all possible messages.

– Dec(s_k, c) decryption: the RSA decryption algorithm generates the plaintext m as follows:

$$m = c^d \bmod n \qquad [8.3]$$

– Homomorphic properties: let $m_1, m_2 \in M$:

$$\begin{aligned} Enc(p_k, m_1) \times Enc(p_k, m_2) &= (m_1 \times m_2)^e \bmod n \\ &= Enc(p_k, m_1 \times m_2) \end{aligned} \qquad [8.4]$$

Clearly, the RSA cryptosystem is homomorphic for the law of multiplication. Indeed, it is possible to calculate $Enc(p_k, m_1 \times m_2)$ from $Enc(p_k, m_1)$ and $Enc(p_k, m_2)$ without decrypting them.

8.2.3.2. *Additive homomorphic encryption*

A scheme (KeyGen, Enc, Dec, Eval) is additively homomorphic if there exists an operation ∘ that verifies the equation:

$$Enc(p_k, m_1) \circ \dots \circ Enc(p_k, m_l) = Enc(p_k, m_1 + \dots + m_l) \qquad [8.5]$$

Additive homomorphic schemes include the GM cryptosystem (Goldwasser and Micali 1982), Benaloh and its derivatives (Benaloh 1994; Naccache and Stern 1998; Okamoto and Uchiyama 1998), and Paillier and its derivatives (Paillier 1999; Damgård and Jurik 2001; Galbraith 2002).

In the following, we describe the GM scheme to explain the principle of an additive HE scheme. This scheme is presented by Goldwasser and Micali (1982) and is defined as follows:

– KeyGen(λ): first, choose two distinct large primes p and q; calculate $n = pq \in \mathbb{Z}$. Then, choose $x \in \mathbb{Z}_n^*$, a quadratic non-residue *modulo* n with $\left(\frac{x}{n}\right) = +1$. Finally, the public key is (x, n) and the private key is (p, q).

– Enc(p_k, m) encryption: given the public key $p_k = (x, n)$ and the plaintext $m \in \{0,1\}$, randomly choose the integer $r \in \mathbb{Z}_n^*$ and calculate the ciphertext as follows:

$$c = r^2 x^m \bmod n \qquad [8.6]$$

Note that this algorithm encrypts only one bit of information, and in all cases $\left(\frac{c}{n}\right) = +1$, but c is a quadratic residue *modulo* n only if $m = 0$.

– Dec(s_k, c) decryption: decryption is equivalent to checking whether c is a quadratic residue *modulo* n or not. If so, the plaintext $m = 0$ or else $m = 1$.

– Homomorphic properties: for each bit $m_i \in \{0,1\}$:

$$\begin{aligned}\mathrm{Enc}(p_k, m_1) \times \mathrm{Enc}(p_k, m_2) &= (r_1^2 x^{m_1} \bmod n) \times (r_2^2 x^{m_2} \bmod n) \\ &= (r_1 \times r_2)^2 x^{m_1+m_2} \bmod n \\ &= \mathrm{Enc}(p_k, m_1 \oplus m_2)\end{aligned} \quad [8.7]$$

Plaintext and ciphertext are elements of the set $\{0,1\}$. The encryption of the sum $\mathrm{Enc}(p_k, m_1 \oplus m_2)$ is calculated from the bits $\mathrm{Enc}(p_k, m_1)$ and $\mathrm{Enc}\ (p_k, , m_2)$ without decrypting them. Consequently, the GM scheme is homomorphic to the XOR operation or *modulo* 2 addition.

Unfortunately, this cryptosystem has a high expansion factor. In fact, a message of l bits are transformed into a ciphertext of $l|n|$ bits ($|n|$: represents the size in bits of the number n). As a result, the expansion factor of the cipher is $l|n\ |/l\ = |n\ |$ bits.

8.2.4. *Somewhat homomorphic encryption*

Somewhat homomorphic encryption (SHE) enables both addition and multiplication. However, one of these operations is limited. In general, multiplication is limited, and addition is unlimited.

Boneh et al. (2005) defined one of the best-known SHE schemes: the BGN scheme (for Boneh Goh Nissim). It allows for an unlimited number of additions and a single multiplication, while maintaining a fixed ciphertext size after homomorphic operations. Unfortunately, this scheme suffers from the large expansion of the ciphertext.

Gentry et al. (2010) improved on the BGN cryptosystem and proposed a scheme that achieves the same homomorphic capabilities, but the encryption/decryption complexities are lower and the ciphertext expansion is also lower.

Herbert et al. (2019) have proposed a variant of the BGN scheme that performs many additions and two multiplications.

The most powerful scheme is the MGH (Melchor Gaborit Herranz) cryptosystem (Melchor et al. 2010), which allows many additions and several multiplications, but the ciphertext grows exponentially by applying homomorphic operations.

8.2.5. *Fully homomorphic encryption*

In his thesis, Gentry (2009a) proposed the first construction of a FHE scheme. This type of scheme allows any function to be evaluated on the encrypted data, and thus performs both types of operation (addition and multiplication) in unlimited numbers. Its security is based on difficult problems in ideal lattices.

8.2.5.1. *The Gentry model*

Gentry's work established not only the first FHE system, but also a paradigm for subsequent FHE schemes. Gentry's construction consists of three steps (Figure 8.3), which can be summarized as follows:

– The first step is to construct a SHE scheme that can evaluate a small number of additions and multiplications. This construction produces a ciphertext of constant size. However, encryption relies on the notion of noise, and so each ciphertext contains a noise parameter that increases exponentially with homomorphic operations. When the noise reaches a certain threshold, the ciphertext becomes undecipherable.

– The second step is to simplify (squash) the decryption circuit as much as possible.

– The final step consists of decrypting the noisy ciphertext and immediately re-encrypting it to reduce the noise. In this procedure, known as bootstrapping, the homomorphic evaluation algorithm evaluates the simplified decryption function on data encrypted using the secret key s_k, encrypted under its own public key (known as the evaluation key). It returns a new actualized ciphertext corresponding to the same plaintext. This results in equation [8.8]:

$$c' \leftarrow Eval\left(p_k, f_{dec}, Enc(p_k, s_k), Enc\left(p_k, c_{noisy}\right)\right) \qquad [8.8]$$

This procedure is the only known method of achieving a pure FHE scheme. However, it is extremely costly and computationally complex, which affects the efficiency of this cryptosystem.

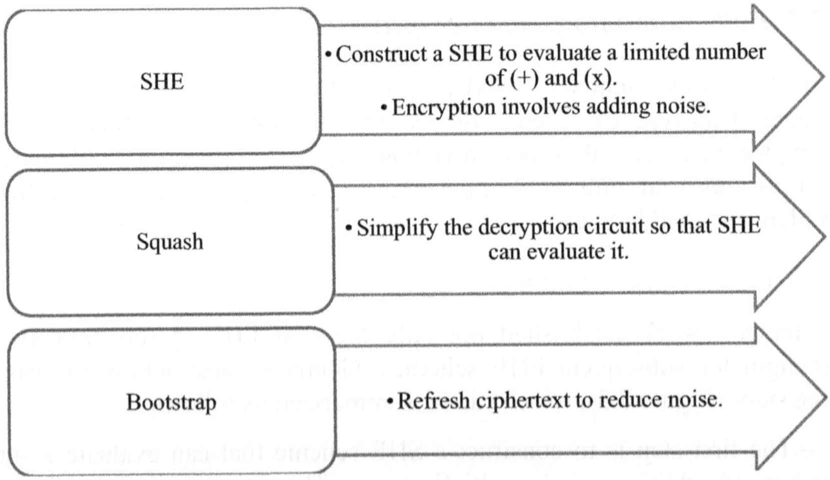

Figure 8.3. *Gentry's model for building an FHE*

8.2.5.2. *Types of FHE schemes*

Following the first FHE cryptosystem introduced by Gentry, the field of HE has undergone considerable expansion. Indeed, many researchers have attempted to design a secure and more efficient FHE scheme. The schemes proposed in the literature are classified into four main families (Acar et al. 2018), according to the mathematical concepts and associated cryptographic problems:

– The first family is based on ideal lattices. It includes the original Gentry (Gen) scheme (Gentry 2009a, 2009b), the Smart Vercauteren (SV) scheme (Smart and Vercauteren 2010) and the Gentry Halev (GH) scheme (Gentry and Halevi 2011). In the SV scheme (Smart and Vercauteren 2010), the authors presented a variant of Gentry's scheme that improves ciphertext expansion and key size. Subsequently, Gentry and Halev (2011) simplified Gentry's key generation procedure and described a better implementation. This class of schemes represents pure FHE schemas.

– The second family concerns FHE schemes over integers. The first scheme was introduced by Van Dijk, Gentry, Halevi and Vaikuntanathan's (vDGHV) (Van Dijk et al. 2010). Its main idea is to apply Gentry's scheme to integers. Therefore, the main interest of this type of scheme is its simple construction, as it uses integer arithmetic instead of ideal lattices. Nevertheless, it suffers from a very large public key size $O(\lambda^{10})$. Coron et al. (2011) reduce this key to $O(\lambda^7)$. Coron et al. (2014) adapt the invariant modulus technique, introduced in Brakerski (2012), to achieve linear ciphertext noise growth.

– The third family of schemes is based on the learning with errors (LWE) problem or the Ring-LWE (R-LWE) problem, which are much simpler problems. The first scheme has been studied by Brakerski and Vaikuntanathan (2011). This class avoids using Gentry's refresh procedure. As a result, the resulting schemes are not actually fully homomorphic, but they do offer significant capacities and greater efficiency. This type of FHE scheme, known as leveled homomorphic encryption (LHE), can evaluate homomorphic operations only for a polynomial of degree L. By increasing the multiplicative depth L, LHE schemes can evaluate any function on encrypted data. Several techniques have been invented to manage and minimize the growth of noise contained in ciphertext after L multiplications. These include tensoring (Brakerski et al. 2012), modulus switching (Brakerski et al. 2012), relinearization (Brakerski 2012; Brakerski et al. 2012), batching (Brakerski et al. 2012) and scale invariance (Brakerski 2012).

– The fourth family of schemes is based on the NTRU cipher, which is an ancient cipher scheme proposed in Hoffstein et al. (1998). The first scheme was created by López-Alt et al. (2012). Their idea is to use multiple keys, making it possible to evaluate several ciphers with different keys. This family is characterized by its efficiency.

8.2.6. *Comparative study*

In the previous sections, we studied the main families of HE schemes: PHE, SHE and FHE. In this section, we present a detailed comparison of some HE schemes with the following criteria:

– Homomorphic capabilities: they represent the set of operations that a scheme can evaluate.

– Ciphertext expansion: it is defined as the ratio of ciphertext size to plaintext size. Many researchers have tried to minimize this ratio.

– Ciphertext size growth: the size of ciphertext increases when homomorphic operations are applied. This problem arises with SHE schemes.

– Based on: it represents the techniques used to design FHE schemes.

– Ciphertext noise growth: this problem arises with FHE schemes. After encryption, the ciphertext contains a noise parameter that will grow with each homomorphic operation. When the noise exceeds a certain threshold, decryption fails.

– Bootstrapping (refreshing): this approach, defined by Gentry, is used to manage the noise contained in the ciphertext and to overcome the problem of ciphertext noise growth. It is the only known method for obtaining a pure FHE scheme, but it is very costly.

We begin by comparing PHE schemes (Table 8.1). These results are taken from (Guellier 2014). We adopt the following notations:

– n: the size of the ciphertext space;

– l: the size of the plaintext in the Benaloh and Naccache–Stern scheme;

– s: an exponent used in the Damgård–Jurik scheme.

Schemes	Operation	Expansion
RSA (Rivest et al 1978)	$\times$	1
Golwasser Micali (Goldwasser and Micali 1982)	$\oplus$	$\|n\|$
ElGamel (ElGamal 1985)	$\times$	2
Benaloh (Benaloh 1994)	$+$	$\|n\|/\|l\|$
Naccache and Stern (Naccache and Stern 1998)	$+$	$\|n\|/\|l\|$
Okamoto and Uchiyama (Okamoto and Uchiyama 1998)	$+$	3
Paillier (Paillier 1999)	$+$	2
Damgård and Jurik (Damgård and Jurik 2001)	$+$	$s/(s+1)$
Galbraith (Galbraith 2002)	$+$	3

Table 8.1. *Comparison of PHE schemes*

Table 8.1 shows that the ElGamel, Paillier and Damgård-Jurik schemes are among the most efficient PHE schemes (in terms of expansion).

A comparison of some SHE schemes in terms of homomorphic capabilities, ciphertext expansion and ciphertext size growth is provided in Table 8.2.

Schemes	Operations	Expansion	Growth in cipher size
BGN (Boneh et al. 2005)	+, × (1)	Large enough	None
GHV (Gentry et al. 2010)	+, × (1)	Smaller than BGN scheme	None
MGH (Melchor et al. 2010)	+, several × ≈polynomial of degree *d*	Large enough	Exponential growth of the figure as a function of d
BGN-F-CF (Herbert et al. 2019)	+, × (2)	Smaller than lattice-based schemes	None

Table 8.2. *Comparison of SHE schemes*

In Table 8.2, we adopt the following acronyms: GHV for Gentry Halevi Vaikuntanathan; and BGN-F-CF for BGN with Freeman transformation and Catalano-Fiore transformation.

Table 8.2 shows that the SHE schemes can be used to evaluate more than one operation (addition and multiplication). The BGN, GHV and BGN-F-CF schemes perform many additions, but with two multiplications, with no growth in ciphertext size; while the MGH scheme evaluates many additions and several multiplications, but at the cost of exponential ciphertext growth. Another important observation is that the smallest expansion of the ciphertext is achieved by the BGN-F-CF scheme (Herbert et al. 2019). It is around 5,000 for $\lambda = 128$ bits (Herbert et al. 2019).

A comparison of several FHE schemes in terms of homomorphic capabilities, ciphertext expansion, ciphertext noise growth and bootstrapping is shown in Table 8.3.

In Table 8.3, we adopt the following acronyms: CMNT for Coron Mandal Naccache Tibouchi; BGV for Brakerski Gentry Vaikuntanathan;

BLLN for Bos Lauter Loftus Naehrig; BV for Brakerski Vaikuntanathan; CLT for Coron Lepoint Tibouchi; KGV for Khedr Gulak Vaikuntanathan; TFHE for fast fully homomorphic encryption over the Torus; and BDF for Bonnoron Ducas Fillinger.

From Table 8.3, we can see that FHE schemes allow an unlimited number of additions and multiplications. Unfortunately, most of these schemes suffer from enormous ciphertext expansion, exponential growth in ciphertext noise and the use of the expensive bootstrapping technique. As a result, FHE schemes are not effective in practice.

After the first FHE scheme, proposed by Gentry (2009a), several schemes were introduced to improve efficiency. For example, in the SV scheme (Smart and Vercauteren 2010), the authors present an FHE scheme with a smaller ciphertext expansion than Gentry's original scheme. In the vDGHV (Van Dijk et al. 2010) and CMNT (Coron et al. 2011) schemes, the authors have established an FHE scheme over integers, resulting in a much simpler scheme than the ideal lattices-based schemes. However, these schemes also suffer from low efficiency. The CLT scheme, proposed by Coron et al. (2014), improved this efficiency. Indeed, the authors used the scale invariance property introduced in Brakerski (2012) to obtain a linear increase in ciphertext noise.

We also conclude from Table 8.3 that LHE schemes are more efficient, but are not, in fact, fully homomorphic. This type of scheme can be transformed into a pure FHE scheme using bootstrapping, but its efficiency becomes comparable to schemes based on ideal lattice. For example, the BGV scheme (Brakerski et al. 2012), which is an LHE scheme, uses the modulus switching technique to efficiently manage ciphertext noise growth. Indeed, this growth becomes polynomial, instead of exponential, for the previous schemes. Bos et al. (2013) proposed the BLLN scheme based on R-LWE. This scheme uses scale invariance and avoids the modulus switching technique. Compared with previous LHE schemes, the BLLN scheme achieves slower growth and better efficiency. In the BV scheme, Brakerski and Vaikuntanathan (2014) introduced the first scheme with linear growth of ciphertext noise. The KGV scheme, also known as SHIELD (Khedr et al. 2016), is an optimized R-LWE variant of the GSW (Gentry Sahai Waters) scheme (Gentry et al. 2013). It features much slower noise growth than previous HE schemes.

Schemes	Based on	Operations	Expansion	Ciphertext growth	Refresh
Gen (Gentry 2009a)	Ideal lattice	+, ×	Yes	Exponential	Yes
SV (Smart and Vercauteren 2010)	Ideal lattice	+, ×	Smaller than Gentry's scheme	Exponential	Yes
vDGHV (Van Dijk et al. 2010)	Integers	+, ×	Yes	Exponential	Yes
CMNT (Coron et al. 2011)	Integers	+, ×	Yes	Exponential	Yes
BGV (Brakerski et al. 2012)	RLWE, LWE	Polynomial of degree L	Little	Polynomial	No
BLLN (Bos et al. 2013)	RLWE	Polynomial of degree L	Little	Polynomial	No
BV (Brakerski and Vaikuntanathan 2014)	LWE	Polynomial of degree L	Little	Linear	No
CLT (Coron et al. 2014)	Integers	Polynomial of degree L	Little	Linear	No
KGV (Khedr et al. 2016)	RLWE	Polynomial of degree L	Little	Linear	No
BDF (Bonnoron et al. 2018)	RLWE	Polynomial of degree L and FHE	Very little	Constant	Yes
TFHE (Chillotti et al. 2020)	RLWE, LWE	Polynomial of degree L and FHE	Very little	Constant	Yes

Table 8.3. *Comparison of FHE schemes*

A very large improvement has been proposed in the BDF (Bonnoron et al. 2018) and TFHE (Chillotti et al. 2020) schemes. First, these improvements introduce methods for speeding up bootstrapping or optimizing the use of bootstrapping. Second, it proposes techniques for packing multiple bits of information and solving the large expansion of ciphertext.

In summary, we can conclude from Tables 8.1–8.3 that most HE schemes trade some efficiency for homomorphic capabilities. PHE schemes perform only one addition or multiplication operation at a time. Their efficiency is close to that of traditional public-key encryption schemes. As for FHE schemes, they enable both addition and multiplication operations. They are theoretically very powerful and are able to compute any function in the encrypted domain. Unfortunately, they are not efficient.

In order to improve this efficiency, two branches of research have been launched. Some research articles focus on reducing bootstrapping time, which is the most expensive part of HE. Ducas and Micciancio (2015) achieve a fast bootstrapping procedure in <1 s. Chillotti et al. (2016) further improve this procedure and achieve it in <0.1 s. More recently, Chillotti et al. (2020) further reduce this time. Other research articles propose to improve LHE schemes and find another alternative to bootstrapping to obtain less complex FHE schemes.

8.2.7. *Application of HE to secure e-health solutions*

HE enables calculations to be performed on encrypted data. In this section, we present possible applications of HE in e-health solutions, to ensure the confidentiality of sensitive medical data.

Bos et al. (2014) propose to use HE to perform predictive analysis on encrypted medical data. The proposed approach is applied as a cloud service for predicting the risk of having a heart attack based on a few body measurements. The service consists of a client application on a private device and a cloud application running in the Microsoft Windows Azure cloud service. The client application collects the user's health data, encrypts it with a LHE scheme and sends the encrypted data to the cloud application, which runs the prediction algorithm based on this data. The cloud produces an encrypted prediction result, which it sends back to the client application.

The user can then decipher and inquire about the probability of having a heart attack. Experiments show that execution time is slow.

An improvement on this proposal was presented in Khedr and Gulak (2018). The authors propose a version of the SHIELD scheme, which reduced its computational complexity by a factor of 4.15. The results of implementing this HE scheme on CPU and GPU show an improvement in computation times compared with the results in Bos et al. (2014). For example, for the cipher multiplication time, the implementation on CPU shows an improvement by a factor of 58. For the same operation, the GPU implementation gives a speed-up of 6,085 and 24,011 using a single GPU and four GPUs respectively.

Medical images contain sensitive patient information. Yin et al. (2020) propose a new medical image encryption algorithm based on elliptic curves, combined with HE.

Similarly, Vengadapurvaja et al. (2017) propose an efficient HE algorithm to encrypt medical images and process them in the cloud without losing their confidentiality. This proposal takes DCOM images as input. To evaluate the performance of this proposal, key space analysis, histogram analysis, correlation analysis, mean square error (MSE), peak signal to noise ratio (PSNR) and noise analysis were performed on medical images. The results of this analysis proved the effectiveness and security of the HE algorithm for medical images.

Personal health monitoring tools, such as wireless ECG patches, can significantly reduce healthcare costs by enabling patients to be monitored outside healthcare facilities. These tools transmit acquired medial data into the cloud, which could be an invaluable diagnostic tool for healthcare professionals. However, there is a risk of losing the confidentiality of patient health information. In Kocabas and Soyata (2020), the authors propose to use HE in order to perform computations on this private health data hosted in the cloud without disclosing the underlying data. The proposed approach is applied as a cloud service for long-term health monitoring. The proposal is tested on three fundamental operations: calculation of average heart rate, calculation of minimum and maximum heart rate, and automatic detection of long QT syndrome. These operations were implemented using the HElib library (Halevi and Shoup 2014), which implements the BGV scheme, and as input, a set of 24 h ECG samples from the THEW database (Couderc 2010). Experimental results show that this solution can be used by healthcare organizations to preserve patient data confidentiality despite its performance issues.

8.3. Attribute-based encryption

ABE is a public-key encryption scheme that uses a single public key for encryption, but which can generate multiple keys for decryption. This is known as one-to-many encryption. It enables access control to encrypted data. The concept of this type of encryption was introduced in 2005 by Sahai and Waters (2005). In this concept, user keys and ciphertext are defined with a set of attributes. A secret key can only decrypt a ciphertext if the cardinality of the intersection of its defined attributes exceeds a fixed threshold. Decryption is therefore only possible for entities with attributes that satisfy an access policy. There are two main types of variant:

– Key-policy attribute-based encryption (KP-ABE), proposed by Goyal et al. (2006): for this type, a trusted authority can generate a secret key associated with an access policy. Only ciphertext with attributes satisfying the private key's access structure can be decrypted.

– Ciphertext-policy attribute-based encryption (CP-ABE), developed by Bethencourt et al. (2007): with this type, each ciphertext is associated with an access policy. A set of attributes is attached to each secret key. Only the secret key with the attributes satisfying the access policy is authorized to decrypt.

An ABE scheme consists of four algorithms:

– Setup(λ): this algorithm takes as input the security parameter λ and returns K_p as a public encryption key and as a secret master key K_M. This K_M key is used to generate secret decryption keys.

– $\mathrm{Enc}(M, K_p, A)$: the encryption algorithm has as input the message M, the public key K_p and a set of attributes A in the case of KP-ABE or an access policy T in the case of CP-ABE. It generates the ciphertext C.

– $\mathrm{KeyGen}(T, K_M)$: the key generation algorithm takes as input the master secret K_M and an access policy T for the KP-ABE case or a set of attributes for the CP-ABE case. It generates a decryption secret key K_s linked to an access policy T for the KP-ABE case or to a set of attributes A for the CP-ABE case.

– $\mathrm{Dec}(K_S, C)$: the decryption algorithm takes as input the user's secret key K_S and the ciphertext C, which has been encrypted under the attribute set. This algorithm produces the message M if and only if the attribute set satisfies the access policy.

A more detailed description of the KP-ABE and CP-ABE variants is given in the following sections.

8.3.1. *Key-policy attribute-based encryption*

In a KP-ABE scheme (Figure 8.4), a trusted authority generates a user's secret key associated with a specific access policy T_i. Ciphertext is generated on the basis of a set of attributes A. This ciphertext can be decrypted by a user if the attributes embedded in C match the access structure specified in their private key.

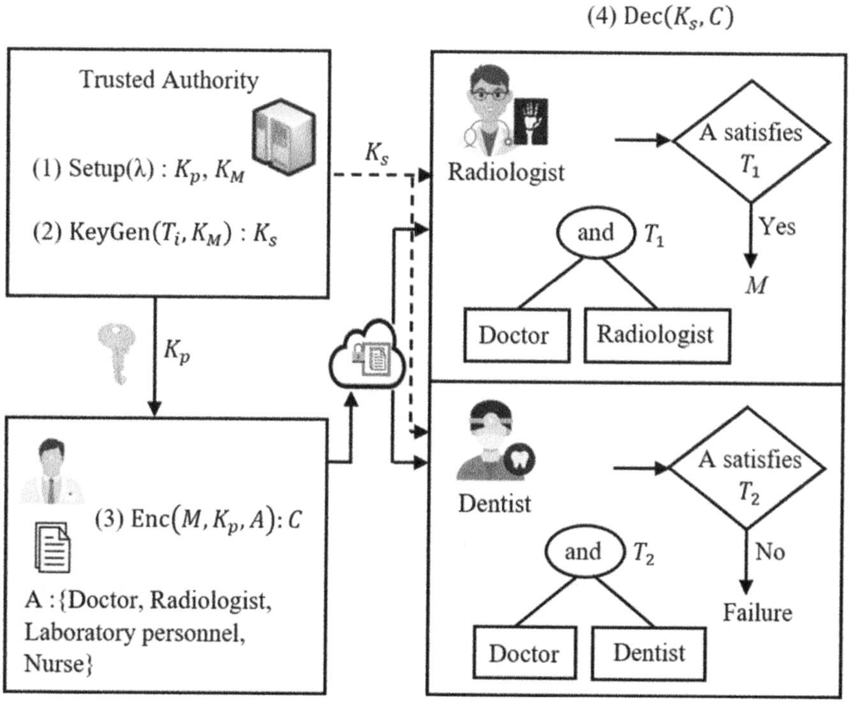

Figure 8.4. *KP-ABE schema*

From Figure 8.4, we can see that the ciphertext, with attributes A = {doctor, radiologist, laboratory personnel, nurse}, satisfies the access structure T_1 of the radiologist user's private key. However, the dentist user cannot decrypt, since the ciphertext with attributes A does not satisfy the access structure T_2 of their secret key.

The KP-ABE variant can achieve fine-grained access control. Its disadvantage is that the data owner cannot decide who can decrypt the ciphertext. They can only choose descriptive attributes for the data. This type of encryption is not suitable for certain applications, as the data owner must trust the key issuer.

8.3.2. *Ciphertext-policy attribute-based encryption*

The CP-ABE variant (Figure 8.5) improves on the disadvantage of the KP-ABE variant. It can support access control in the real environment. In addition, the user's private key in this scheme is a combination of a set of attributes A_i. So, a user must use only this set of attributes to satisfy the access structure T in the ciphertext.

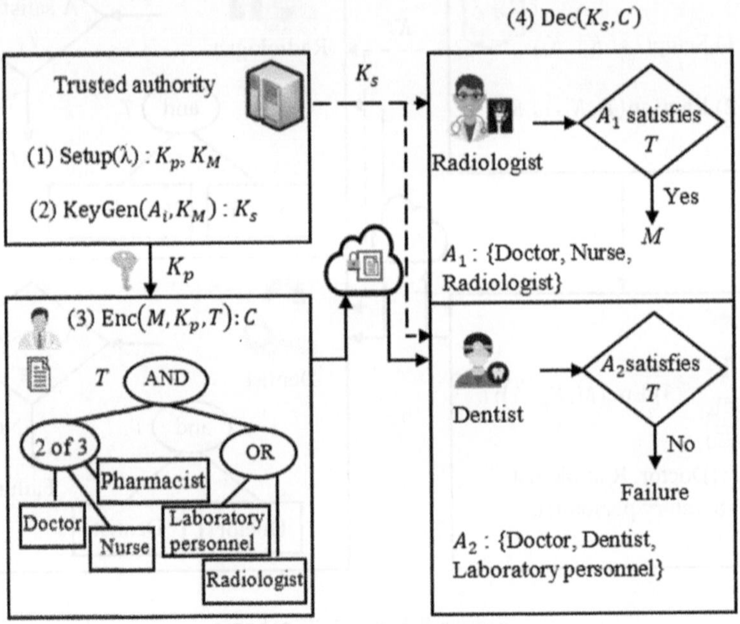

Figure 8.5. *CP-ABE schema*

From Figure 8.5, we can see that access policy T is embedded in the ciphertext, and attributes A_1 are used to generate the radiologist's private key. This key, with attributes A_1 = {doctor, nurse, radiologist}, satisfies access policy T, so the radiologist can retrieve message M, whereas the dentist's private key, with attributes A_2 = {doctor, dentist, laboratory personnel}, does not satisfy access policy T. The dentist is unable to decrypt the ciphertext.

KP-ABE	CP-ABE
A set of attributes is attached to each ciphertext.	A set of attributes is attached to each private key.
The private key is associated with an access policy that controls the ciphertext a user can decrypt.	Ciphertext is associated with an access policy that controls which user can decrypt the ciphertext.
The owner of the data cannot decide who can decrypt the ciphertext.	The owner of the data can decide who can decrypt the ciphertext.
This type of encryption is not suitable for certain applications, as the data owner must trust the key issuer.	It supports access control in the real environment.

Table 8.4. *Main differences between KP-ABE and CP-ABE*

The CP-ABE variant has limitations in terms of policy specification and user attribute management. In addition, decryption keys only support user attributes that are logically organized in a single set. Therefore, users can only use all possible combinations of attributes in a single set issued in their keys to satisfy policies.

Table 8.4 shows the main differences between the KP-ABE and CP-ABE schemes.

8.3.3. *Comparative study*

In Table 8.5, we compare encryption schemes by attributes by the following criteria:

– Data confidentiality: unauthorized parties cannot learn about encrypted data in the cloud, including the cloud server.

– Fine-grained access: the system must grant different access rights to different users, even those in the same group.

– User accountability: when an authorized user shares their key with unauthorized users, he/she must be held individually accountable.

– Scalability: when the number of authorized users increases, system performance must not be affected.

– User revocation: when users leave the system, they cannot access any stored data.

Schemes	Confidentiality	Fine-grained access	User accountability	Scalability	User revocation
KP-ABE	Yes	Yes	No	No	Yes
CP-ABE	Yes	Yes	Yes	No	Yes

Table 8.5. *Comparison of ABE schemes*

8.3.4. *Application of ABE to secure e-health solutions*

With the aim of achieving confidentiality, scalability and fine-grained access to data outsourced to the cloud, Yu et al. (2010) propose a scheme combining KP-ABE attribute encryption, proxy re-encryption (Blaze et al. 1998) and lazy re-encryption (Kallahalla et al. 2003). In addition, this proposed scheme allows the data owner to delegate most complex computational operations to cloud servers without disclosing data content or user access privilege information. The formal security analysis shows that the proposed scheme is efficient and secure under existing security models.

To ensure the confidentiality of personal health information (PHI) hosted in the cloud, Barua et al. (2011) propose an Efficient and Secure Patient-centric Access Control (ESPAC) scheme. The proposed scheme uses CP-ABE. It enables data requesters to have different access privileges depending on their roles (doctor, pharmacist, etc.). Security and performance analysis results have shown that the proposed scheme is highly efficient in resisting a variety of possible attacks and malicious behavior.

The work presented in Hupperich et al. (2012) proposes an approach based on ABE attribute encryption to protect the confidentiality of electronic health records (EHRs). The authors use ABE and public key encryption for scalable secret authorization. The patient's smart card generates a Transaction Access Code (TAC) that represents the authorization secret.

This code can be sent to a doctor by phone or other remote mechanism. Encryption of medical data with a public key by the healthcare professional is only possible after verification of the TAC authorization code. Decryption requires the physician to possess the TAC authorization code and to acquire authentication from the private key generator.

Qian et al. (2014) present a study on key management, fine-grained access control and fast user revocation. They demonstrated fine-grained access restriction on personal health records (PHRs), while protecting privacy.

To ensure the security of PHRs hosted in the cloud, Yan et al. (2014) propose a system using the KP-ABE technique. This approach enables fine-grained access control and greater efficiency for user revocation.

Gowda et al. (2017) propose a new solution to ensure the confidentiality of EHR medical records outsourced in the cloud. This solution uses hierarchical ABE and secure keyword search. It encrypts electronic medical records by integrating hierarchical access structures. This system guarantees confidentiality, while enabling fine-grained access control and flexible user revocation.

Sangeetha and Vaidehi (2017) present a system in which patients can encrypt their PHRs and place them on cloud servers. They can restrict access to their PHRs by providing fine-grained access rights based on attributes. The suggested PHR structure is separated into personal domains (PSD) and public domains (PUD). Privacy-preserving ABE algorithms are applied in PSDs and PUDs, respectively, to ensure security in a cloud-based PHR architecture.

Tembhare et al. (2019) propose a system that integrates both RBAC (role-based access control) and ABE schemes. This system attempts to ensure PHR access while respecting patient privacy.

Gardiyawasam Pussewalage and Oleshchuk (2022) propose an efficient access control system based on CP-ABE. This proposition enables its users fine-grained access to externalized EHR electronic medical records in the cloud and allows users to delegate access to externalized files in a controlled manner when required. Security analysis results have shown that the proposed scheme is secure against selected plaintext attacks.

References	Type	Proposition	Access control
(Yu et al. 2010)	KP-ABE	Secure access to outsourced data in the cloud	Fine grained
(Barua et al. 2011)	CP-ABE	Confidentiality of personal health information (PHI)	Role based
(Hupperich et al. 2012)	ABE	Confidentiality of EHR files	Scalable
(Qian et al. 2014)	CP-ABE	Respecting the privacy of the PHR file	Fine grained
(Yan et al. 2014)	KP-ABE	Secure access to PHR	Fine grained
(Gowda and Sumathi 2017)	CP-ABE	Secure EHR sharing	Fine grained
(Sangeetha and Vaidehi 2017)	KP-ABE	PHR cloud security	Fine grained
(Tembhare et al. 2019)	KP-ABE	Protecting the privacy of PHR files	Role based
(Gardiyawasam Pussewalage and Oleshchuk 2022)	CP-ABE	Secure access to EHR files	Fine grained

Table 8.6. *Summary of ABE solutions for securing e-health*

A summary of the application of ABE in securing e-health solutions is given in Table 8.6.

8.4. Conclusion

Most medical data processed in the cloud contains sensitive and strictly confidential patient information. Consequently, data confidentiality is a major challenge in cloud computing.

In this chapter, we look at how HE and ABE can help secure e-health solutions. In the first part of this chapter, we studied the basics of HE encryption. Then, we described its different families: PHE, SHE and FHE. This study enabled us to compare several HE schemes in terms of homomorphic capabilities, ciphertext expansion, ciphertext noise growth and bootstrapping. Finally, we studied the contribution of HE to securing e-health solutions. In the second part of this chapter, we studied the basics of ABE encryption. We then studied and compared its different variants,

KP-ABE and CP-ABE. Finally, we present examples of how ABE can be applied to secure e-health solutions.

8.5. References

Acar, A., Aksu, H., Uluagac, A.S., Conti, M. (2018). A survey on homomorphic encryption schemes: Theory and implementation. *ACM Computing Surveys*, 51(4), 1–35.

Barua, M., Liang, X., Lu, R., Shen, X. (2011). ESPAC: Enabling security and patient-centric access control for e-health in cloud computing. *International Journal of Security and Networks*, 6(2/3), 67–76.

Benaloh, J. (1994). Dense probabilistic encryption. In *Proceedings of the Workshop on Selected Areas of Cryptography*. Springer, Kingston, 120–128.

Bethencourt, J. and Waters, B. (2007). Ciphertext-policy attribute-based encryption. In *2007 IEEE Symposium on Security and Privacy (SP'07)*, Oakland, 321–334.

Blaze, M., Bleumer, G., Strauss, M. (1998). Divertible protocols and atomic proxy cryptography. In *International Conference on the Theory and Applications of Cryptographic Techniques*. Springer, Espoo, 127–144.

Boneh, D., Goh, E.J., Nissim, K. (2005). Evaluating 2-DNF formulas on ciphertexts. In *Theory of Cryptography Conference*. Springer, Cambridge, MA, 325–341.

Bonnoron, G., Ducas, L., Fillinger, M. (2018). Large FHE gates from tensored homomorphic accumulator. In *International Conference on Cryptology in Africa*. Springer, Marrakesh, 217–251.

Bos, J.W., Lauter, K., Loftus, J., Naehrig, M. (2013). Improved security for a ring-based fully homomorphic encryption scheme. In *IMA International Conference on Cryptography and Coding*. Springer, Oxford, 45–64.

Bos, J.W., Lauter, K., Naehrig, M. (2014). Private predictive analysis on encrypted medical data. *Journal of Biomedical Informatics*, 50, 234–243.

Brakerski, Z. (2012). Fully homomorphic encryption without modulus switching from classical GapSVP. In *Annual Cryptology Conference*. Springer, Santa Barbara, CA, 868–886.

Brakerski, Z. and Vaikuntanathan, V. (2011). Fully homomorphic encryption from ring-LWE and security for key dependent messages. In *Annual Cryptology Conference*. Springer, Santa Barbara, CA, 505–524.

Brakerski, Z. and Vaikuntanathan, V. (2014). Efficient fully homomorphic encryption from (standard) LWE. *SIAM Journal on Computing*, 43(2), 831–871.

Brakerski, Z., Gentry, C., Vaikuntanathan, V. (2012). (Leveled) fully homomorphic encryption without bootstrapping. In *ITCS: Proceedings of the 3rd Innovations in Theoretical Computer Science Conference*. Association for Computing Machinery, New York, 309–325.

Chillotti, I., Gama, N., Georgieva, M., Izabachène, M. (2016). Faster fully homomorphic encryption: Bootstrapping in less than 0.1 seconds. In *International Conference on the Theory and Application of Cryptology and Information Security*. Springer, Hanoi, 3–33.

Chillotti, I., Gama, N., Georgieva, M., Izabachène, M. (2020). TFHE: Fast fully homo-morphic encryption over the torus. *Journal of Cryptology*, 33(1), 34–91.

Coron, J.S., Mandal, A., Naccache, D., Tibouchi, M. (2011). Fully homomorphic encryption over the integers with shorter public keys. In *Annual Cryptology Conference*. Springer, Santa Barbara, CA, 487–504.

Coron, J.S., Lepoint, T., Tibouchi, M. (2014). Scale invariant fully homomorphic encryption over the integers. In *International Workshop on Public Key Cryptography*. Springer, Buenos Aires, 311–328.

Couderc, J.P. (2010). The Telemetric and Holter ECG Warehouse initiative (THEW): A data repository for the design, implementation and validation of ECG-related technologies. In *2010 Annual International Conference of the IEEE Engineering in Medicine and Biology Society, EMBC'10*, Buenos Aires, 6252–6255.

Damgård, I. and Jurik, M. (2001). A generalisation, a simplification and some applications of Paillier's probabilistic public-key system. In *International Workshop on Public Key Cryptography*. Springer, Jeju Island, 13–15.

Ducas, L. and Micciancio, D. (2015). FHEW: Bootstrapping homomorphic encryption in less than a second. In *Annual International Conference on the Theory and Applications of Cryptographic Techniques*. Springer, Sofia, 617–640.

ElGamal, T. (1985). A public key cryptosystem and a signature scheme based on discrete logarithms. *IEEE Transactions on Information Theory*, 31(4), 469–472.

Galbraith, S.D. (2002). Elliptic curve Paillier schemes. *Journal of Cryptology*, 15(2), 129–138.

Gardiyawasam Pussewalage, H.S. and Oleshchuk, V. (2022). A delegatable attribute based encryption scheme for a collaborative e-health cloud. *IEEE Transactions on Services Computing*, 1374(c), 1–14.

Gentry, C. (2009a). A fully homomorphic encryption scheme. PhD Thesis, Stanford University, Stanford.

Gentry, C. (2009b). Fully homomorphic encryption using ideal lattices. In *Proceedings of the Forty-First Annual ACM Symposium on Theory of Computing*, Bethesda, MD, 169–178.

Gentry, C. and Halevi, S. (2011). Implementing Gentry's fully-homomorphic encryption scheme. In *Annual International Conference on the Theory and Applications of Cryptographic Techniques*. Springer, Tallinn, 129–148.

Gentry, C., Halevi, S., Vaikuntanathan, V. (2010). A simple BGN-type cryptosystem from LWE. In *Annual International Conference on the Theory and Applications of Cryptographic Techniques*. Springer, French Riviera, 506–522.

Gentry, C., Sahai, A., Waters, B. (2013). Homomorphic encryption from learning with errors: Conceptually-simpler, asymptotically-faster, attribute-based. In *Annual Cryptology Conference*. Springer, Santa Barbara, CA, 75–92.

Goldwasser, S. and Micali, S. (1982). Probabilistic encryption & how to play mental poker keeping secret all partial information. In *Proceedings of the Fourteenth Annual ACM Symposium on Theory of Computing*, San Francisco, CA, 173–201.

Gowda, B.K. and Sumathi, R. (2017). Hierarchy attribute-based encryption with timing enabled privacy preserving keyword search mechanism for e-health clouds. In *2nd IEEE International Conference on Recent Trends in Electronics, Information & Communication Technology (RTEICT)*, Bangalore, 425–429.

Goyal, V., Pandey, O., Sahai, A., Waters, B. (2006). Attribute-based encryption for fine-grained access control of encrypted data. In *Proceedings of the 13th ACM Conference on Computer and Communications Security*, Alexandria, VA, 89–98.

Guellier, A. (2014). Can homomorphic cryptography ensure privacy? PhD Thesis, Supélec Rennes, Rennes.

Halevi, S. and Shoup, V. (2014). GitHub – shaih/HElib: An Implementation of homomorphic encryption [Online]. Available at: https://github.com/shaih/HElib [Accessed 3 July 2022].

Herbert, V., Biswas, B., Fontaine, C. (2019). Design and implementation of low-depth pairing-based homomorphic encryption scheme. *Journal of Cryptographic Engineering*, 9(2), 185–201.

Hoffstein, J., Pipher, J., Silverman, J.H. (1998). NTRU: A ring-based public key cryp-tosystem. In *International Algorithmic Number Theory Symposium*. Springer, Portland, OR, 267–288.

Hupperich, T., Löhr, H., Sadeghi, A.R., Winandy, M. (2012). Flexible patient-controlled security for electronic health records. In *Proceedings of the 2nd ACM SIGHIT International Health Informatics Symposium*, New York, 727–732.

Kallahalla, M., Riedel, E., Swaminathan, R., Wang, Q., Fu, K. (2003). Plutus: Scalable secure file sharing on untrusted storage. In *2nd USENIX Conference on File and Storage Technologies (FAST 03)*, San Francisco, CA, 29–42.

Khedr, A. and Gulak, G. (2018). SecureMed: Secure medical computation using GPU-accelerated homomorphic encryption scheme. *IEEE Journal of Biomedical and Health Informatics*, 22(2), 597–606.

Khedr, A., Gulak, G., Vaikuntanathan, V. (2016). SHIELD: Scalable homomorphic implementation of encrypted data-classifiers. *IEEE Transactions on Computers*, 65(9), 2848–2858.

Kocabas, O. and Soyata, T. (2020). Towards privacy-preserving medical cloud computing using homomorphic encryption. In *Virtual and Mobile Healthcare: Breakthroughs in Research Practice*, Information Resources Management Association (ed.). IGI Global, Hershey, PA, 93–125.

López-Alt, A., Tromer, E., Vaikuntanathan Mit, V. (2012). On-the-fly multiparty computation on the cloud via multikey fully homomorphic encryption. In *Proceedings of the Forty-Fourth Annual ACM Symposium on Theory of Computing*, New York, 1219–1234.

Melchor, C.A., Gaborit, P., Herranz, J. (2010). Additively homomorphic encryption with d-operand multiplications. In *Annual Cryptology Conference*. Springer, Santa Barbara, CA, 138–154.

Naccache, D. and Stern, J. (1998). A new public key cryptosystem based on higher residues. In *ACM Conference on Computer and Communications Security*, San Francisco, CA, 59–66.

Okamoto, T. and Uchiyama, S. (1998). A new public-key cryptosystem as secure as factoring. In *International Conference on the Theory and Applications of Cryptographic Techniques*. Springer, Espoo, 308–318.

Paillier, P. (1999). Public-key cryptosystems based on composite degree residuosity classes. In *International Conference on the Theory and Applications of Cryptographic Techniques*. Springer, Prague, 223–238.

Qian, H., Li, J., Zhang, Y., Han, J. (2014). Privacy-preserving personal health record using multi-authority attribute-based encryption with revocation. *International Journal of Information Security*, 14(6), 487–97.

Rivest, R.L., Adleman, L., Dertouzos, M.L. (1978). On data banks and privacy homomorphisms. *Foundations of Secure Computation*, 4(11), 169–180.

Rothblum, R. (2011). Homomorphic encryption: From private-key to public-key. In *Theory of Cryptography Conference*. Springer, Providence, RI, 219–234.

Sahai, A. and Waters, B. (2005). Fuzzy identity-based encryption. In *Annual International Conference on the Theory and Applications of Cryptographic Techniques*. Springer, Aarhus, 457–473.

Sangeetha, D. and Vaidehi, V. (2017). A secure cloud based personal health record framework for a multi owner environment. *Annals of Telecommunications*, 72(1/2), 95–104.

Smart, N.P. and Vercauteren, F. (2010). Fully homomorphic encryption with relatively small key and ciphertext sizes. In *International Workshop on Public Key Cryptography*. Springer, Paris, 420–443.

Tembhare, A., Sibi Chakkaravarthy, S., Sangeetha, D., Vaidehi, V., Venkata Rathnam, M. (2019). Role-based policy to maintain privacy of patient health records in cloud. *Journal of Supercomputing*, 75(9), 5866–5881.

Van Dijk, M., Gentry, C., Halevi, S., Vaikuntanathan, V. (2010). Fully homomorphic encryption over the integers. In *Annual International Conference on the Theory and Applications of Cryptographic Techniques*. Springer, French Riviera, 24–43.

Vengadapurvaja, A.M., Nisha, G., Aarthy, R., Sasikaladevi, N. (2017). An efficient homomorphic medical image encryption algorithm for cloud storage security. *Procedia Computer Science*, 115, 643–650.

Yan, H., Li, X., Du, J. (2014). Secure personal health record system with attribute-based encryption in cloud computing. In *9th International Conference on P2P, Parallel, Grid, Cloud and Internet Computing, 3PGCIC 2014*. IEEE, Guangdong, 329–332.

Yin, S., Liu, J., Teng, L. (2020). Improved elliptic curve cryptography with homomorphic encryption for medical image encryption. *Int. J. Netw. Secur.*, 22(3), 419–424.

Yu, S., Wang, C., Ren, K., Lou, W. (2010). Achieving secure, scalable, and fine-grained data access control in cloud computing. In *Proceedings – IEEE INFOCOM*, San Diego, CA, 1–9.

9

Artificial Intelligence for Security of e-Health Systems

Mohamed Aymen CHALOUF[1], Hana MEJRI[2] and Omessaad HAMDI[3]
[1] *IRISA, Université de Rennes, France*
[2] *Sys'Com-ENIT, Université de Tunis El Manar, Tunis, Tunisia*
[3] *IEEE, Rennes, France*

9.1. Introduction

The Internet of Things (IoT) has grown considerably in recent years. This development has been fueled by major advances in information and communication technologies (ICT), which have enabled the democratization of connected objects and accelerated the deployment of IoT applications. This has been facilitated by the widespread deployment of new access technologies, such as low power wide area networks (LPWAN), and the reduction in the cost of communicating objects.

The majority of IoT applications are focused on improving users' daily lives and increasing productivity in a number of fields, such as industry and agriculture. Among the fields that have benefited from technological advances and IoT development, we find healthcare. Indeed, electronic health, also known as e-health, is one of the areas propelled forward because of the various capabilities for collecting, analyzing, transmitting, storing and making available medical data made possible with the development of the IoT.

e-Health Security Management,
coordinated by Omessaad HAMDI.

In this chapter, we focus on the use of artificial intelligence (AI) techniques to secure one of the main components of e-health systems, namely the wireless medical sensor network known as wireless body area network (WBAN). In section 9.2, we describe the architecture and the components of e-health systems. Section 9.3 introduces the security challenges of e-health systems, as well as the solutions and techniques used to meet these challenges. Then, section 9.4 describes the main AI techniques used to provide one of the most important security services within the WBAN, namely intrusion detection. After that, section 9.5 details the use of AI techniques for intrusion detection, before section 9.6 focuses on AI-based intrusion detection systems (IDSs) in WBANs. Finally, section 9.7 concludes this chapter.

9.2. e-Health systems

The high cost of caring for the elderly and patients with reduced autonomy or chronic illnesses in hospitals and nursing homes has prompted the development and deployment of alternative solutions. Initially, these alternative solutions made it possible to keep elderly people and patients in their own homes, while guaranteeing them the necessary health monitoring and care. Later, these solutions were extended to include other healthcare services, such as remote diagnosis, which revolutionized the healthcare world during the Covid-19 pandemic. These solutions are mainly based on ICT to enable healthcare professionals, such as doctors and nurses, to remotely monitor patients' health states and provide them with the needed care. These solutions also enable healthcare professionals, and even patients, to access health data. The first component of e-health is telemedicine, covering four acts (Simon 2010): remote consultation, remote expertise, remote care and remote monitoring.

To guarantee e-health services, e-health systems have been implemented. The study of these systems, detailed in Hamdi et al. (2014), identified a general architecture that has been detailed in the first chapter of this book. This architecture is made up of three main parts:

– A wireless medical sensors network: it is composed of heterogeneous wireless medical sensors that equip the patient. These sensors take measurements and collect data to provide healthcare professionals with information on the patient's health state. Examples include body temperature sensors, heart rate monitors, electrocardiograms and SpO2 (pulse oxygen saturation) sensors, which measure oxygen levels in the blood.

– A gateway: it is an intermediate node between the patient's local network and the external network. This gateway may be a medical box, or a smartphone when the patient is at home or on the move, respectively.

– An external network: it consisted of the rest of the infrastructure of an e-health system. This part involves powerful calculation and storage servers and enables the gateway to transmit the patient's medical data to them. These servers act as the link between the patient and the healthcare professionals, and have the capacity to make some decisions, such as alerting a doctor or a relative in case of an emergency.

In an e-health system, several types of communication are possible. First, communications within the wireless WBAN. Next, communications between the WBAN and the gateway. Then, communications between the gateway and remote servers on the external network. Finally, communications between the remote servers and the terminals of other actors, such as healthcare professionals. These communications can be wired or wireless.

In addition, medical data are not only exchanged between the different parts of an e-health system but can also be processed and/or stored on equipment in the patient's local network, such as a sensor, smartphone or medial box. These data can also be processed and/or stored on external network equipment, such as remote servers, which are characterized by much greater capacities than those of the patient's local network equipment.

e-Health systems offer a wide range of advantages, making it easier to care for patients and improving their daily lives and living conditions. However, when an e-health system is targeted by cyber-attacks, this can lead to catastrophic situations. In section 9.3, we look at the security of e-health systems.

9.3. e-Health system security

In an e-health system, the falsification or the redirection of an alert could have serious consequences on a patient's health state, up to and including death. Given the sensitive and private nature of medical data transiting, processed and stored in e-health systems, their protection must be ensured.

9.3.1. *Potential attacks*

The architecture of an e-health system, introduced in the first chapter of this work, distinguishes two main networks: (i) the patient's local network including the WBAN and the gateway; (ii) the external network including remote servers. Equipment on these networks may be involved in the processing, storage and transmission of private, sensitive and critical medical data. The multilocation of data storage and its exchange via wireless and mobile networks increase the attack surface of e-health systems. Attacks are thus manifold and can threaten the physical security of equipment, or the security of storage and communications.

Generally, security attacks fall into two categories: passive and active (Chalouf 2020). Active attacks allow unauthorized access to information, with or without modification. In e-health systems, we find other classifications. For example, we can distinguish between attacks linked to data storage and access (Ming et al. 2010) and those linked to data exchange (Shahnaz et al. 2011). Furthermore, another classification (Shihong et al. 2017) distinguishes between attacks targeting the local network, that is, the WBAN (intra-WBAN), and those targeting exchanges between the local network and the external network (extra-WBAN). For exchanges between the local network and the external network, we find the classic attacks, such as passive eavesdropping, forgery, identity theft, replay and denial of service (DoS). With regard to attacks targeting the local area network (WBAN), three different types can be distinguished (Javadi et al. 2013). First, attacks threatening confidentiality and authentication, where the attacker performs eavesdropping, replay or identity theft. Second, integrity-threatening attacks aimed at forcing the WBAN to accept false information. Finally, attacks threatening unavailability, where the attacker aims to disable the WBAN.

9.3.2. *Security services*

Security attacks on e-health systems can cause significant damage to patients. Therefore, these systems need to be secured by protecting the data processed, exchanged, and stored within them. Security needs can be divided into three classes (Chalouf 2020): privacy and data access, data exchange and data storage.

9.3.2.1. *Privacy and data access*

The first class of services concerns privacy and secure data access:

– Confidentiality protects patient data against leakage and disclosure. To protect patient privacy, all communications in an e-health system must be encrypted.

– Access control prevents unauthorized persons or entities from accessing patient data. To secure access to data, role-based access control techniques must be implemented.

– Responsibility allows us to identify and hold accountable a user who abuses his or her rights by performing unauthorized actions on patient data.

– Revocability protects against compromised users and nodes. If a user/node is compromised, it must be stripped of all previously granted authorizations.

– Non-repudiation ensures that the sender of a message cannot deny being the source of that message, and that the recipient cannot deny receipt of the message.

9.3.2.2. *Communications security*

The second class of services concerns the security of communications:

– Confidentiality: to guarantee the confidentiality of patient medical data as these are exchanged between different parts of the e-health system, the data must be encrypted at source and decrypted at destination.

– Data integrity: to guarantee that the data received have not been altered, it is necessary to be able to verify that the data received are authentic and correct.

– Data origin authentication: to prevent an illegitimate entity from masquerading as a legitimate one, it is necessary to be able to verify the identity of the data sender.

– Data freshness: to ensure the freshness of data received, it is necessary to be able to check that data frames are received in the right order and are not replayed.

– Data availability: to ensure that patient medical data are available to authorized persons, it is necessary to be able to resist DoS attacks.

9.3.2.3. *Storage security*

The third class of services concerns storage security:

– Confidentiality: patient medical data must remain confidential not only when they are exchanged, but also when they are stored.

– Integrity: the integrity of patient medical data must be verified not only when they are exchanged, but also when they are stored.

– Reliability: patient medical data must be recoverable, even in the event of a problem or failure of part of the e-health system.

9.3.3. *Security solutions*

Security services are provided by deploying security mechanisms and techniques. Table 9.1 (Chalouf 2020) shows some examples of the mechanisms and techniques deployed to secure WBANs.

Security attacks	Security services	Security mechanisms
Unauthorized data access	Access control and trust management	– Authentication protocols – Reputation management
Disclosure of data	Confidentiality and privacy	– Data encryption – Access control
Data modification	Integrity and authentication	– HMAC, digital signature
Denial of service	Availability	– Intrusion detection, redundancy
Compromised node	Resistance to node compromise	– Inconsistency detection – Node revocation
Intrusion	Intrusion detection	– Intrusion detection system

Table 9.1. *Possible solutions for WBAN security (Chalouf 2020)*

Data exchanged within the local network (intra-WBAN) and between the local network and the external network (extra-WBAN) must be protected: confidentiality, integrity, authentication of origin, non-repudiation, etc. Confidentiality is ensured by symmetrical, asymmetrical or elliptic curve encryption. Authentication of origin and integrity are jointly guaranteed by HMAC (Hash Message Authentication Code) or a digital signature, which also guarantees non-repudiation. Data freshness and processing in the right order can be ensured by adding a sequence number. Access control is ensured by simple (single factor) or strong (multi-factor) access rules and authentication mechanisms.

The security of extra-WBAN exchanges is generally ensured using classic security protocols, such as IPSec (Frankel et al. 2011), TLS (Rescola 2018) or DTLS (Rescola et al. 2012). As for the security of intra-WBAN exchanges, this requires security mechanisms adapted to the specificities of this network. For example, an access control protocol for medical data that respects patient privacy has been proposed in Sammoud et al. (2020b). Also, a solution for managing the responsibility of users of patient medical data was proposed in Yu et al. (2009). In addition, solutions to ensure data freshness (Saleem et al. 2009) have been proposed. Otherwise, solutions have been proposed to ensure availability, such as those based on redundancy (Javadi et al. 2013), enabling the transfer of functionality from a disabled or compromised node to another available node.

Patient medical data are often stored on external servers, such as those in the cloud. The security of this storage is ensured by encrypting the data prior to storage. This security can be enhanced, and patient privacy respected through mechanisms such as anonymization (Li et al. 2010a), or unlinkability, between medical data and the patient and confidentiality of the patient identifier (Ermakova et al. 2013).

Some security services, such as integrity, origin authentication and confidentiality, are provided by symmetric or asymmetric cryptography. Symmetric cryptography has the advantage of speed and low resource consumption (processor time, memory and energy). However, it requires efficient and secure key distribution. As for asymmetric cryptography, it has the advantage of robustness and simple key distribution. Nevertheless, it is not suited to the security of intra-WBAN communications (Vijayalakshmi 2011), given the limited capabilities of sensors (Carman et al. 2000). Thus, the local network of an e-health system is often secured using symmetrical cryptography, while at the same time proposing new techniques for symmetrical key management, such as those based on biometric data. As an example, we cite the protocol proposed in Sammoud et al. (2020a) and discussed in Chapter 6 of this book.

In addition, an e-health system must be able to authenticate nodes and users to control access to sensitive medical data. Given the specificities of e-health systems, such as the constraints of some equipment, user mobility and the availability of certain biometric data, new multifactor authentication solutions based on biometric data and adapted to e-health systems have been proposed. An example of such a solution (Sammoud et al. 2020b) is given in Chapter 7 of this book.

One of the essential security services for protecting e-health systems is intrusion detection at local network (intra-WBAN). Intrusion detection is generally provided by an IDS, which detects abnormal or suspicious activity on a network or device. This system identifies successful or failed intrusion attempts on the basis of signatures or anomalies. Anomaly-based detection involves identifying deviations in the behavior of the target system (network or equipment) and is often based on AI techniques, notably machine learning.

In the remainder of this chapter, we focus on intrusion detection in the local network of an e-health system (intra-WBAN) using AI techniques. Section 9.4 introduces the main AI techniques used in this security service (intrusion detection).

9.4. Artificial intelligence techniques

AI is a set of technologies that enable us to reproduce human intelligence, or at least come close to it. AI is at the heart of progress in various fields, such as healthcare and e-health. In healthcare in general, AI and its various techniques, such as machine learning, are transforming the way healthcare is delivered. Indeed, AI technologies are well-suited to different treatments and analyses of medical data, with the aim of making better decisions, particularly in terms of diagnosis and treatment.

In e-health systems, AI can be used not only to implement security services, such as intrusion detection, but also to carry out processing with the aim of making effective decisions for the control and management of e-health system security. For example, AI can be used to improve the effectiveness of the proposed framework for protecting IoT applications (Sylla et al. 2019) by ensuring automatic and dynamic implementation of protection adapted to the global context of any IoT application, including e-health applications. Indeed, AI, and more specifically machine learning, would make it possible to assess the robustness of the protection mechanisms decided by the proposed framework and their convenance to each context in order to improve decision-making in terms of the security mechanisms to implement. Furthermore, the new multifactor authentication protocols based, among other things, on biometric data and involving a patient or healthcare professional (Sammoud et al. 2020b, 2020c) can be strengthened by defining new multimodal biometric authentication mechanisms that will make use of AI techniques. Indeed, these techniques will enable the detection of the living character and thus achieve more secure authentication.

AI techniques used for intrusion detection are essentially learning based on two main categories: machine learning and deep learning.

9.4.1. *Machine learning*

Machine learning takes raw data and automatically transforms it into knowledge, with the aim of making the best decisions. It can discover patterns and make predictions from the data at hand, based on statistics, classification and data partitioning. Machine learning thus makes it possible to exploit massive, structured and categorized data, to understand these data and to be able to process new, similar information so as to make the right decisions.

Machine learning (surface learning) techniques are divided into two types: supervised learning and unsupervised learning.

9.4.1.1. *Supervised learning*

Supervised learning is a machine-learning approach that defines a prediction function from labeled data sets. This labeled data represents a learning base and is designed to enable algorithms to classify data and predict results accurately. This assumes that the learning base is representative, and supervised learning methods are able to generalize predictions to data not present in the learning base. By using labeled inputs and outputs, the model (learned prediction function) can measure its accuracy and learn over time. Supervised learning algorithms are used to accurately assign test data to specific categories. For example, they can be used to classify spam in a separate folder in the e-mail inbox. Depending on the nature of the variable being predicted, two problems can be distinguished: regression and classification. The prediction of a qualitative variable is considered a classification problem. Examples of classification algorithms are linear classifiers, support vector machines, decision trees and decision tree forests. As for the prediction of a quantitative variable, this is considered a regression problem. In fact, regression is a supervised learning method that uses algorithms to understand the relationship between dependent and independent variables. There are three types of regression: linear regression, logistic regression and polynomial regression.

9.4.1.2. *Unsupervised learning*

Unsupervised learning is a machine-learning approach for defining a prediction function from a set of unlabeled data. The techniques of this approach make it possible to analyze and cluster sets of unlabeled data in order to discover hidden patterns in the data without the need for human intervention. These patterns are usually represented by probability densities or preference combinations of neuronal features and are generally capable of reaching decisions without any assistance or intervention. Generally, unsupervised learning techniques can perform more complex tasks than supervised learning techniques. However, they can be unpredictable and produce results that can disrupt the prediction system. Non-supervised learning algorithms must adapt to unknown data and situations, and their decisions can sometimes be less accurate than those of supervised learning algorithms. These algorithms are mainly based on data partitioning, enabling different sets of data to be distinguished where each of these sets has common characteristics.

In conclusion, the aim of machine learning is to learn how to classify new data. Two approaches can be distinguished: supervised learning and unsupervised learning. The aim of supervised learning is to learn to identify the class of new data from a set of predefined classes. These classes are already known and identified, and this is the very purpose of human supervision and intervention. However, unsupervised learning aims to learn, by itself, classes or groups of data sharing some characteristics. The number and definition of these classes are not known in advance, and this is the meaning of unsupervised learning. In this case, the quality of the classification is measured by the ability to discover "patterns".

9.4.2. *Deep learning*

Deep learning represents another approach to learning. This approach is very different from machine learning, as it does not require structured data to operate. A deep learning system is capable of learning reliable classification models using incomplete data. This system is widely used for complex tasks for which no structured data is available upstream and does not need to be trained beforehand to be able to make a decision or reproduce a precise task.

A deep learning system is organized into several processing layers, forming neural networks that combine different algorithms inspired by the human brain. Each layer will categorize the input information, according to precise, predefined criteria and decide on a classification for this information, which will be found in the output data. Deep learning requires a greater quantity of data than machine learning and consumes considerable resources.

9.5. Intrusion detection based on artificial intelligence

There are many different types of attack that can target e-health systems from the local network (intra-WBAN) to the external network (extra-WBAN). Attacks targeting the local network include intrusion at host level (a WBAN node or gateway) or at network level (WBAN). These attacks must be detected, and countermeasures implemented to protect the local network of an e-health system. Detection is generally based on an IDS, which will detect any abnormal or suspicious activity on the potential target (host or network) being monitored and analyzed. For example, network traffic monitoring using specific software is essential for network intrusion detection. In 1980, J.P. Anderson was the first to demonstrate the importance of security auditing in detecting violations of a system's security policy. He defined an intrusion as "a deliberate and unauthorized attempt to access or modify information, or to render a system unusable or inaccessible" (Anderson 1980, authors' translation).

There are two types of IDS: signature based and anomaly based. Signature-based detection uses a signature database that needs to be updated regularly in order to adapt to and detect newly identified attacks. This type of IDS has the advantage of being effective against some categories of intrusion, such as malware, and producing a low number of false positives. However, it can only detect known attacks (signatures present in the database), making it vulnerable to unknown attacks (such as *zerodays*). Anomaly-based detection, on the other hand, can identify a wider range of intrusion attacks by basing its detection on deviations in behavior relative to a model representing good behavior. This second category of detection (anomaly-based) often uses learning techniques. Indeed, given the ever-increasing number of attacks targeting networks, IDSs have become indispensable. In this context, many research have been carried out, and learning-based detection solutions have emerged (Kumar et al. 2010). The main objective of these solutions is to detect known and unknown attackers (intrusions and intrusion attempts) efficiently and accurately, while minimizing the number of false positives (Bailey et al. 2017). AI-based IDSs can implement the techniques introduced in section 9.4: supervised learning, unsupervised learning or deep learning.

Unlike signature based IDSs, IDSs implementing supervised learning can detect certain unlisted attacks. To do so, they must first define the normal behavior of the protected network. Normal behavior is defined through modeling, which enables the IDS to be trained and the normality model to be obtained. Detection is then carried out by evaluating the deviation of any event or behavior from the normal state (defined by the model). If this deviation exceeds a certain threshold, the activity is qualified as abnormal.

9.5.1. *IDS based on supervised learning*

IDSs based on supervised learning use statistical or classification techniques. The first category (statistical) defines an anomaly as suspicious behavior and performs tests on data from a statistical model to decide whether new behavior corresponds to an anomaly or not. For example, some IDSs in this category take into account the statistical properties of network flows, such as the duration and latency of a flow, or the frequency and length of packets in that flow. These IDSs do not need to know about normal activity in advance and are effective when the monitoring period is long enough, and the alert threshold is well calibrated. The second category (classification) assigns a behavior to a class. It performs automatic learning on pre-classified data to obtain rules for classifying unknown behaviors. These IDSs therefore need the pre-classified data to establish the model. Classification can be linear or nonlinear. Among the classification-based techniques used in IDS are the k-nearest neighbor method, the decision tree and the support vector machine. These techniques are characterized by their flexibility, enabling them to adapt to changes, and their good detection efficiency, especially for known attacks.

However, IDSs based on supervised learning have a number of drawbacks. For example, IDSs implementing statistical techniques require a long time to build the model, and their effectiveness depends on the statistics used and the parameterization impacting the number of false positives and false negatives. Also, these IDSs can be misled by an attacker. Furthermore, the effectiveness of IDSs based on classification techniques is highly dependent on classification assumptions and the relevance of training data.

Unlike IDSs based on statistics and classification, IDSs based on data partitioning offer more stable performance. Data partitioning is the main technique used in IDSs based on unsupervised learning.

9.5.2. *IDS based on unsupervised learning*

IDSs implementing unsupervised learning have been proposed to improve the detection of unclassified attacks. These IDSs provide more reliable detection of unknown attacks, including polymorphic attacks. These IDSs are also able to adapt to changes in the network or protected system. IDSs based on unsupervised learning often use the data partitioning technique. This technique divides data into several sets, where each set groups together data with common characteristics. In these IDSs, when learning fails to assign a behavior, event or traffic to a known data set, the latter is termed abnormal. For optimum performance, an IDS based on unsupervised learning needs to have a dataset comprising many more normal behaviors than abnormal behaviors. Abnormal behavior data must have very different characteristics from normal behavior data. Also, data corresponding to abnormal behaviors must not be classified in a group with a similar size to those containing data representing normal behaviors. Otherwise, the abnormal behaviors corresponding to these data may be considered normal.

There are several unsupervised learning techniques based on data partitioning. IDSs use some of these techniques by implementing algorithms, such as the k-means clustering algorithm or hierarchical clustering algorithms. These techniques are characterized by good reliability, especially when the data corresponding to threats and abnormal behavior have one or more different characteristics. The response time of these IDSs can be reduced by using a supervised learning mechanism to perform an initial classification of the data, before subsequently analyzing the data with unsupervised learning.

The reliability of these IDSs depends not only on the degree of similarity between normal and abnormal data, but also on the quantity of abnormal data. At the same time, these techniques can be very resource-intensive, given the complexity of identifying some anomalies and threats. As with IDSs based on supervised learning, the ability of IDSs based on unsupervised learning to adapt to change remains limited. This adaptation to change can be improved with IDSs based on deep learning.

9.5.3. *IDS based on deep learning*

IDSs based on deep learning can detect intrusions and threats more reliably and faster than IDSs based on superficial learning (supervised or unsupervised). This is due to the ability of these IDSs to analyze data in depth. Indeed, deep learning builds the classification model based on incomplete data by precisely specifying the characteristics of the anomalous data. Classification is achieved by calculating the distance separating the different normalized data. Deep learning uses algorithms that simulate the functioning of the human brain, with neural networks made up of several successive layers of neurons. As a result, IDSs implementing deep learning have managed to outperform other IDSs (surface learning) and are now capable of accurately detecting a greater number of attacks.

However, this detection reliability depends not only on the quality of the data used in the learning process, such as the level of noise (amount of meaningless or corrupted data), but also on the quantity of these data. Indeed, IDS based on deep learning work well when the quantity of training data is large. This implies another advantage of this category of IDS (deep learning), whose efficiency increases with increasing data quantity, whereas that of the other category of IDS (shallow learning) reaches a saturation point after a certain time. Nevertheless, the complexity of deep learning algorithms and the need for large amounts of data imply a significant consumption of resources, which must be taken into consideration when choosing the IDS to implement in a given context.

9.6. AI-based IDS in WBANs

Since each context of use (application flows, network architecture, access technology, equipment, and devices, etc.) has its own characteristics, one AI technique may be more suitable than another one for intrusion detection in a given context. "Adapted" does not necessarily mean more efficient, as other criteria may need to be taken into account, such as resource consumption in a constrained environment, like WBANs, which are composed of sensors with limited computational and energy capacities. Furthermore, in a WBAN, communications are wireless, and the radio environment is often disrupted by sources of interference that can increase the percentage of erroneous messages and, consequently, of corrupted data used in the learning process on which the IDS is based. If this noise is not tolerated, it can lead to an increase in the IDS false-positive rate and reduce its effectiveness. As some techniques can tolerate noise more than others, the aim of the work described in this section is to implement and compare several AI techniques in order to identify the one that best tolerates noise and best adapts to a disturbed radio environment (noise).

9.6.1. *Tested learning techniques*

In this work, we compared three supervised learning techniques based on classification:

– Decision trees: the decision tree is a supervised learning algorithm used to perform classification and prediction on a data set. Typically, it starts with a single node and, based on this node, linked to the algorithm's input variable, generates several nodes representing all possible outcomes. A decision tree comprises three types of nodes: decision nodes, chance nodes and end nodes.

– Neural networks: artificial neural networks are networks of highly connected elementary processors operating in parallel. Each elementary processor calculates a single output on the basis of the information received as input. Generally, a neural network is characterized by parallelism (several basic elementary entities working simultaneously), adaptability (self-organization to adapt to new constraints and data) and generalizability (data not included in training data), as well as fault tolerance (local failures at the level of some elements).

– Bayesian networks: naive Bayesian classifiers are a set of classification algorithms based on Bayes' theorem. Bayes' theorem determines the probability of an event occurring, given the probability that another event has already occurred. These linear classifiers implement simple probabilistic Bayesian classification based on feature independence.

9.6.2. *Implementation and results*

The learning techniques tested can be used to detect anomalies (intrusion or attempted intrusion) within the local network of an e-health system. The implementation and evaluation of each of these techniques includes different phases:

– Data acquisition and pre-processing (cleaning): the data obtained must be sufficient and representative of the problem to be solved. The data must then be processed to ensure that only strictly interesting data are retained. This will result in a more accurate model, optimizing its size, as well as learning and execution times.

– Model realization: model realization consists in finding the best structure, as well as the set of initialization parameters. The complexity and quality of the model have an impact on the accuracy of data generalization.

– Training: a subset of the cleaned data forms the training data, enabling the model's learning phase to be carried out. This enables the model parameters to be adjusted.

– Validation: during this phase, the model and its parameters are tested and validated according to a set of criteria. This allows us to obtain the best model generalizing the data obtained during the learning phase.

– Evaluation: a model is evaluated by comparing its performance with that of other models on the same dataset. Performance indicators include model size, learning time, execution time and percentages of true positives, true negatives, false negatives and false positives.

To carry out the tests, we used a dataset relating to intrusion attacks. The dataset used is the WSN-DS (Almomani et al. 2016), which consists of eighteen attributes and five possible outputs: four types of attack (anomalies) and a normal state (no anomaly and, therefore, no attack). WSN-DS contains 374,661 records, with 340,066 normal records. The rest of the records are divided between the four types of attack considered (blackhole, grayhole, flooding and scheduling).

These data are quite structured. However, some of them need a little transformation, so that they can be used in the modeling phase. After modeling, models are evaluated to find the model that best represents the dataset under consideration. These data have been divided into two parts. The first part comprises the examples used to prepare the model (training), representing 66% of the dataset. The second part consists of examples used to test and evaluate the learning techniques considered.

The first evaluation of the models obtained with the three implemented learning techniques yielded the results shown in Figures 9.1–9.3. These results show a detection accuracy of 99% for the decision tree-based IDS, a detection accuracy of 94% for the neural network-based IDS and a detection accuracy of 95% for the naive Bayesian classification-based IDS.

As a reminder, detection precision (A: accuracy) is calculated as follows:

$$A = \frac{TPR + TNR}{TPR + FPR + TNR + FNR}$$

where TPR, TNR, FPR and FNR represent, respectively, the true positive rate, true negative rate, false positive rate and false negative rate, with, for example:

$$TPR = \frac{TP}{TP + FN} \qquad TNR = \frac{TN}{TN + FP}$$

We also recall that each confusion matrix summarizes the results of intrusion attack prediction using the classification-based supervised learning techniques we have considered. These matrices compare real data with the data predicted by each model, where one row corresponds to a real class and one column to a predicted class.

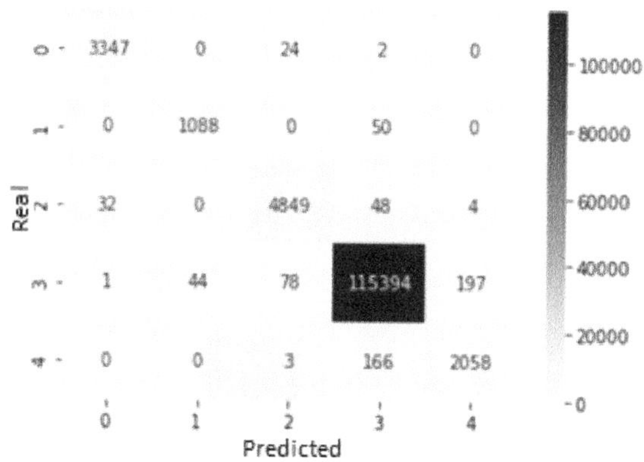

Figure 9.1. *Confusion matrix for the "decision tree" model*

Given the interference that can affect communications in wireless medical sensor networks (WBANs), we set out to assess the impact of this interference on the performance of IDSs based on AI techniques. We injected random noise into the data used for detection. We also varied the level of noise injected from 5% to 40% (5%, 10%, 20%, 30% and 40%). This is equivalent to generating different datasets (with noise) from the original dataset.

As an example, Figures 9.4 and 9.5 show the results obtained with an IDS implementing the decision tree algorithm with the following noise levels: 5% and 40%.

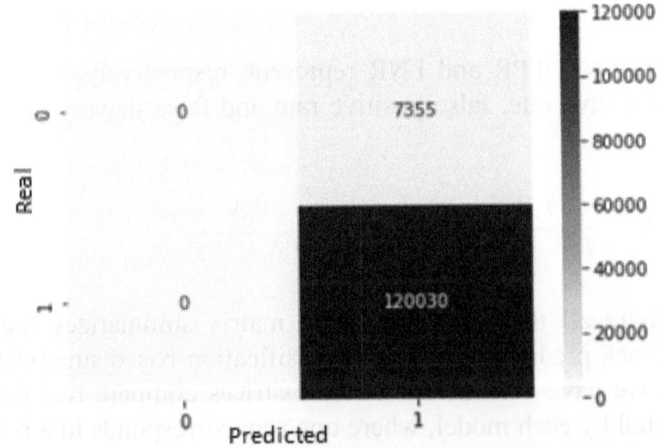

Figure 9.2. *Confusion matrix for the neural network model*

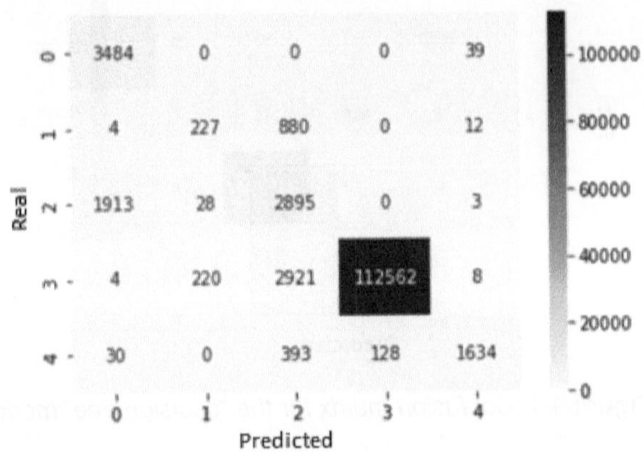

Figure 9.3. *Confusion matrix for the "naive Bayes" model*

These results confirm the noise sensitivity of this IDS, with an impact that increases as the noise level rises. For example, the number of true positives for the normal class rose from 115,188 to 114,937, and the number of false positives for this class rose from 330, when the noise level was 5%, to 675 when it rose to 40%.

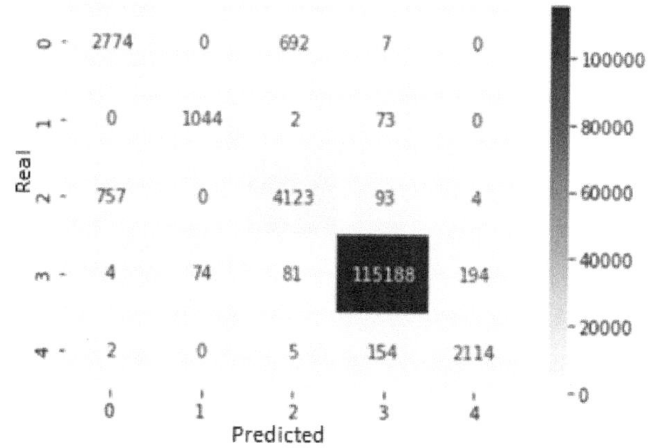

Figure 9.4. *Confusion matrix for the "decision tree" model with 5% noise*

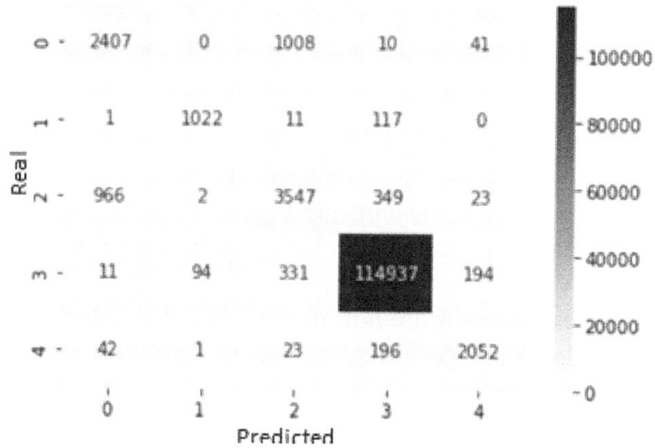

Figure 9.5. *Confusion matrix for the "decision tree" model with 40% noise*

Evaluation of the performance of IDSs based on the three AI techniques considered (decision tree (DT), artificial neural network (ANN) and Bayesian networks (naive Bayes)), in an environment characterized by interference (noise), shows that the IDS based on neural networks is the technique that best tolerates such interference (see Figure 9.6).

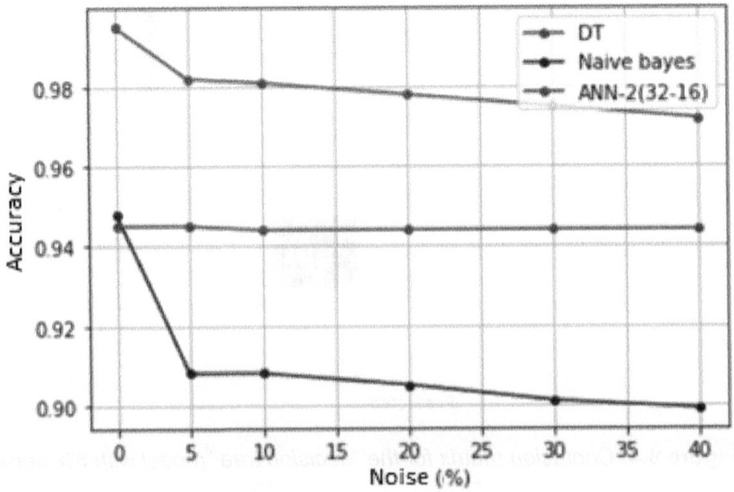

Figure 9.6. *IDS performance as a function of noise level. For a color version of this figure, see www.iste.co.uk/hamdi/ehealth.zip*

9.7. Conclusion

The first e-health systems appeared several years ago. Today, these systems have become a reality and have really contributed to improving patients' daily lives, while lowering the cost of certain treatments and follow-ups. As these systems are not immune to cyber-attacks, it has become essential to develop high-performance security solutions adapted to these systems. Among the security services to be provided, we were interested in intrusion detection at the level of the patient's local network (intra-WBAN). Our aim was to implement and test a number of IDSs based on AI techniques. We then subjected these IDSs to different levels of noise in order to assess their ability to tolerate this noise, which can be very significant, particularly when the patient is mobile. The results shown in this chapter were obtained by randomly injecting noise. Our next objective is to inject noise according to models that are closer to reality and to evaluate the noise tolerance threshold of the techniques evaluated, which can be extended to other AI techniques. We will also be able to study the consumption of resources and the response time of IDSs. The results of this study will enable us to select the method best suited to the capabilities of the sensors forming the WBAN. When intrusion detection is provided by a distributed IDS (DIDS), detection nodes need to exchange detection data. Federated learning can be used to limit these exchanges and avoid some related security problems (Lavaur et al. 2022). In this technique, training is carried out on the local data of the nodes, which just need to establish a scheme for exchanging predictive models and for aggregating and comparing the results obtained by the different nodes.

9.8. References

Almomani, I., Al-Kasasbeh, B., Al-Akhras, M. (2016). WSN-DS: A dataset for intrusion detection systems in wireless sensor networks. *Journal of Sensors*, 2, 1–16. doi: 10.1155/2016/4731953.

Anderson, J.P. (1980). Computer security threat monitoring and surveillance. Report, Early NIST Security Papers.

Bailey, M., Collins, C., Sinda, M., Hu, G. (2017). Intrusion detection using clustering of network traffic flows. In *18th IEEE/ACIS International Conference on Software Engineering, Artificial Intelligence, Networking and Parallel/Distributed Computing (SNPD)*. doi: 10.1109/snpd.2017.8022786.

Carman, D.W., Kruus, P.S., Matt, B.J. (2000). Constraints and approaches for distributed sensor network security. NAI Labs Technical Report.

Chalouf, M.A. (2020). Managing quality of service and security in an e-health environment. In *Service Level Management in Emerging Environments*, Mbarek, N. (ed.). ISTE Ltd, London, and John Wiley & Sons, New York.

Ermakova, T. and Benjamin, F. (2013). Secret sharing for health data in multi-provider. In *IEEE 15th Conference on Business Informatics*, Vienna, 93–100. doi: 10.1109/CBI.2013.22.

Frankel, S. and Krishnan, S. (2011). IP security (IPsec) and Internet Key Exchange (IKE) document roadmap. Internet Engineering Task Force (IETF).

Hamdi, O., Chalouf, M.A., Ouattara, D., Krief, F. (2014). e-Health: Survey on research projects, comparative study of telemonitoring architectures and main issues. *Journal of Network and Computer Applications*, 46, 100–112.

Javadi, S.S. and Razzaque, M.A. (2013). *Security and Privacy in Wireless Body Area Networks for Health Care Applications, Wireless Networks and Security, SCT*. Springer, Berlin.

Kumar, G., Kumar, K., Sachdeva, M. (2010). The use of artificial intelligence based techniques for intrusion detection: A review. *Artificial Intelligence Review*, 34, 369–387. doi: 10.1007/s10462-010-9179-5.

Lavaur, L., Pahl, M.O., Busnel, Y., Autrel, F. (2022). The evolution of federated learning-based intrusion detection and mitigation: A survey. *IEEE Transactions on Network and Service Management*, 19(3), 2309–2332.

Li, M., Yu, S., Ren, K., Lou, W. (2010a). *Securing Personal Health Records in Cloud Computing: Patient-Centric and Fine-Grained Data Access Control in Multi-owner Settings, Security and Privacy in Communication Networks*. Springer, Berlin.

Li, M., Lou, W., Ren, K. (2010b). Data security and privacy in wireless body area networks. *IEEE Wireless Communications*, 17(1), 51–58. doi: 10.1109/MWC.2010.5416350.

Rescorla, E. (2018). The Transport Layer Security (TLS) Protocol Version 1.3. Internet Engineering Task Force (IETF). Request for Comments.

Rescorla, E. and Modadugu, N. (2012). Datagram Transport Layer Security Version 1.2. Internet Engineering Task Force (IETF). Request for Comments.

Saleem, S., Ullah, S., Yoo, H.S. (2009). On the security issues in wireless body area networks. *International Journal of Digital Content Technology and its Applications*, 3, 178–184. doi: 10.4156/jdcta.vol3.issue3.22.

Sammoud, A., Chalouf, M.A., Hamdi, O., Montavont, N., Bouallègue, A. (2020a). A new biometrics-based key establishment protocol in WBAN: Energy efficiency and security robustness analysis. *COSE*, 15, 96. doi: 10.1016/j.cose.2020.101838.

Sammoud, A., Chalouf, M.A., Hamdi, O., Montavont, N., Bouallègue, A. (2020b). A secure and lightweight three-factor authentication and key generation scheme for direct communication between healthcare professionals and patient's WMSN. In *25th IEEE Symposium on Computers and Communications, ISCC 2020*, Rennes.

Sammoud, A., Chalouf, M.A., Hamdi, O., Montavont, N., Bouallègue, A. (2020c). A secure three-factor authentication and biometrics-based key agreement scheme for TMIS with user anonymity. In *16th International Conference on Wireless Communications & Mobile Computing, IWCMC 2020*, Limassol.

Shahnaz, S., Sana, U., Kyung, S.K. (2011). A study of IEEE 802.15.4 security framework for wireless body area networks. *Sensors*. 11, 1383–1395. doi: 10.3390/s110201383.

Shihong, Z., Yanhong, X., Honggang, W., Zhouzhou, L., Shanzhi, C., Bo, H. (2017). A survey on secure wireless body area networks. *Journal of Security and Communication Networks*, 9. doi: 10.1155/2017/3721234.

Simon, P. (2010). New organizations and healthcare management with telemedicine. *Médecine des maladies métaboliques*, 4, 251–256. doi: 10.1016/S1957-2557(10)70056-0.

Sylla, T., Chalouf, M.A., Krief, F., Samaké, K. (2019). Towards a context-aware security and privacy as a service in the Internet of Things. In *13th WISTP International Conference on Information Security Theory and Practice, WISTP 2019*, Paris.

Vijayalakshmi, B. (2011). A zero-knowledge authentication for wireless sensor net-works based on congruence. In *3rd IEEE International Conference on Advanced Computing*, Chennai, 94–99. doi: 10.1109/ICoAC.2011.6165155.

Yu, S., Ren, K., Lou, W., Li, J. (2009). Defending against key abuse attacks in KP-ABE enabled broadcast systems. In *5th International ICST Conference on the Security and Privacy in Communication Networks*. SecureComm, Athens, 311–329.

List of Authors

Ryma ABASSI
Innov'COM
SUP'COM
Université de Carthage
Tunis
Tunisia

Manel ABDELHEDI
Noccs
ENISo
Université de Sousse
Tunisia

Mohamed AYMEN CHALOUF
IRISA
Université de Rennes
France

Aida BEN CHEHIDA DOUSS
Innov'COM
SUP'COM
Université de Carthage
Tunis
Tunisia

Faiza HAMDI
Jeddah University
Saudi Arabia

Omessaad HAMDI
IEEE
Rennes
France

Francine KRIEF
LaBRI
Bordeaux INP
France

Cyrine LAHSINI
Sys'Com-ENIT
Université de Tunis El Manar
Tunis
Tunisia

Hana MEJRI
Sys'Com-ENIT
Université de Tunis El Manar
Tunis
Tunisia

Léo MENDIBOURE
ERENA
COSYS
Université Gustave Eiffel
Marne-la-Vallée
France

Amal SAMMOUD
MIRA
Aflokkat
Ajaccio
France

Tidiane SYLLA
GEII
ISA
USTTB
Bamako
Mali

Index

A

access control, 13, 15, 18, 19, 58, 59, 67, 68, 70, 72–74, 78, 83, 92, 100, 129, 168, 200, 202–205, 217–219
architecture(s), 28, 43, 46, 68, 72, 74, 84, 86, 89–94, 100, 109, 131, 137, 138, 162, 205, 214, 216, 226
artificial intelligence, 8, 49, 81, 100, 126, 127, 214, 220, 221, 223, 224, 226, 229, 231, 232

B

BAN (Body Area Network), *see also* WBAN, 70, 137–139, 146, 162, 172, 214
biometrics, 14, 16, 17, 68, 137, 141, 145–147, 154, 156, 161, 163, 166, 168
blockchain, 14, 18, 19, 45, 49, 73, 79, 89–94, 96–98, 100, 107–133

C, D

cloud, 6, 8, 10–12, 15, 17, 28, 30, 31, 42, 43, 45, 73, 80, 162, 183–185, 198, 199, 203–206, 219
  computing, 28, 43, 183–186, 206
cyber-attack, 20, 30, 34, 40, 41
decentralization, 45, 112, 127, 131

E, I

electrocardiogram (ECG), 15, 38, 70, 84, 87, 137, 143–150, 154–157, 166, 167, 169, 173, 175, 176, 178, 199

encryption, 14–16, 18–20, 44, 45, 67–70, 124, 126, 141, 147, 154, 171, 184–193, 198–200, 202–206, 218, 219
attribute-based, 184, 185, 202–205
homomorphic, 45, 184, 185, 187, 188, 192, 193, 198, 199, 206
interoperability, 72, 108, 119–121, 125–127, 130, 131, 133

M, P

monitoring/follow-up, 29, 35, 37, 44, 68, 69, 84, 120, 121, 124, 161, 214
penetration(s)/intrusion(s), 32, 42, 46, 214, 218, 220, 221, 223, 226–229, 232
permission/authorization, 33, 45, 62, 63, 72, 73, 78, 79, 87, 91–97, 99, 100, 142, 204

S, T, W

sensor(s), 84, 87, 142, 144, 146, 179, 215
traceability, 112, 115–119, 121, 123–125
WBAN (Wireless Body Area Network), *see also* BAN, 137–148, 150, 154, 155, 162, 214–216, 218–220, 223, 226, 229, 232

Printed and bound by CPI Group (UK) Ltd, Croydon, CR0 4YY
13/09/2026
14961070-0002